BUYINC

Buying the Big Jets
Fleet Planning for Airlines
Second Edition

PAUL CLARK

ASHGATE

Published by
Ashgate Publishing Limited
Gower House
Croft Road
Aldershot
Hampshire GU11 3HR
England

Ashgate Publishing Company
Suite 420
101 Cherry Street
Burlington, VT 05401-4405
USA

Ashgate website: http://www.ashgate.com

British Library Cataloguing in Publication Data
Clark, Paul, 1954-
  Buying the big jets : fleet planning for airlines
  1. Airlines - Planning 2. Airplanes - Purchasing
  3. Airplanes - Evaluation
  I. Title
  387.7'3'0687

Library of Congress Cataloging-in-Publication Data
Clark, Paul, 1954-
  Buying the big jets : fleet planning for airlines / by Paul Clark.
        p. cm.
  "2nd edition"--Publisher's summary.
  Includes index.
  ISBN 978-0-7546-7090-2 (hardback)
  ISBN 978-0-7546-7091-9 (pbk.)
  1. Airlines--Management. 2. Airlines--Planning. 3. Airplanes--Purchasing. I.
Title.

  HE9780.C55 2007
  387.7068'4--dc22

2006036920

ISBN: 978-0-7546-7090-2 (hardback)
ISBN: 978-0-7546-7091-9 (pbk.)

Reprinted 2008

Printed and bound in Great Britain by MPG Books Ltd, Bodmin, Cornwall.

Contents

List of Figures

List of Tables

Preface to the Second Edition

Since the first edition of this book appeared in 2001, some radical changes have occurred in airline fleet planning.

Firstly, the emergence of the low-cost carriers has had an impact on the way aircraft are marketed and analysed. Low-cost operations are successful thanks to stringent, if not ruthless, application of their one guiding principle; simplicity of product and operation. As a result the aircraft is little more than a commodity expected to deliver efficiency. However, the notion of the aircraft as an airline product differentiator is still very much alive and well for network and long-haul carriers. So, a second development in fleet planning has been the resurgence of the cabin interior in the high-stakes game of airline market domination. The conflict of least-cost simplification *versus* higher cost customisation is a much bigger fleet planning issue than before. It is a clash of ideologies.

Another tendency is that the task of the fleet planner has become more and more one of project management. He or she is now the conductor of a vast orchestra of talented players, who may not always be playing the same tune, or even be playing *in* tune.

Finally, the aircraft manufacturers are engaged in playing the final movement in their own symphonies. Product families are being rounded-out and completed with the battle lines now clear for all to see.

This book addresses the processes of fleet planning as applied to scheduled airlines operating aircraft with a capacity of 100 seats or more. We will follow the typical steps, starting with how fleet planning is organised, looking at how to measure market behaviour, assessing and defining the aircraft product, following the analytical processes of performance and economics, and concluding with a look at the investment appraisal and the final strategic decision.

In the First Edition I asked the question, 'Is fleet planning an art or a science?' I still have absolutely no idea.

Acknowledgements

When Ashgate suggested that a Second Edition of *Buying the Big Jets* might be a good idea, I must admit that I froze. The nightmares that had accompanied the writing of the original book over five years earlier had not entirely faded. Of course, I knew that Ashgate were right, and thank them, in particular Guy Loft and Sarah Cooke, for their encouragement. So, after innumerable feeble excuses for extending the deadline, I finally succumbed and launched myself into the project of updating the book.

It is often said that writing is a solitary exercise. It is not. Producing a textbook of this nature is only possible if the writer is prepared to browbeat his friends and colleagues for data, advice, encouragement, reassurance and, on occasion, a spot of amateur psychotherapy and pots of hot tea. My victims acted with generosity and grace without exception.

In particular I would like to thank Christine Lebied, who took the task of checking my writing on aircraft performance and economics so seriously that I actually felt guilty. Claude Pluzanski returned from a three-week vacation and instantly replied to my requests for data. Anaïs Marzo, with whom I have travelled around the world so many times to run seminars on air transport, generously shared her own experiences and data. Luckily, Anaïs is aware of most of my faults so was able to 'orient' my thinking. Marianne Way, Librarian at AirBusiness Academy, was always on call to provide data at the drop of a hat. Nils Hallerstrom of PK AirFinance declared that he would be proud if I used two charts prepared by himself and Jan Melgaard. Nils mistakenly believes that publication of the charts will hasten his Nobel Prize.

This book is a personal project, and not at all linked to my employer. This allows me to include the time-honoured phrase; all views expressed are those of the author. However, I must thank Airbus for their support. I also thank IATA, whose data are unsurpassed in interest and value.

In addition to the willing contributors listed above, there is another group of friends and colleagues who have made equally significant contributions to this book. Dick Wyatt, Tim Jeans, David Stewart, Stephen Shaw, Didier Lenormand, Jonathan Lesieur, Andrew Gordon, Rick Jones, Professor Roger Wootton, Louis Busuttil, Eef van Herpt, Dr Adam Pilarski, Mireille Mielvaque, Xavier Burgat, Jean-Pierre Stainnack, Sean Jackson, Pierre Suteau, Philippe Jarry, Simon Pickup, Joseph Debacq, William Gibson and Professor Rigas Doganis have all unwittingly shaped my thinking during various discussions over the course of the last year. I must also thank my personal assistant at AirBusiness Academy, Soraya Hachemi, for her professionalism and efficiency. By 'managing' my daily life in the office she achieved the virtually impossible task of putting me in the right frame of mind to write this book.

That this book exists at all is thanks to the people who truly inspire me. Judith, Christopher and Robin graciously accepted my troglodyte existence during the writing phase and I thank them for their amazing forbearance. Erin Egan and Richard Olivier have also been an inspirational force and provided a much-needed spark just when it was really needed.

I dedicate this book to the memory of Philippe Tournoux, the very first marketing training professional at Airbus, who died in an aircraft accident. Had he lived, he would have written this book.

Chapter 1

The Big Picture

What, Exactly, is Fleet Planning?

The best laid schemes o' mice an' men
Gang aft a-gley
An' lea'e us nought but grief an' pain
For promis'd joy[1]

Robert Burns

The eighteenth century Scottish poet Robbie Burns penned these words in 1786. He was telling us that no matter how much effort we make in planning, there is always something lurking around the corner to throw our ideas into disarray. Naturally, we cannot predict the unpredictable. However, we ought to be able to predict the consequences of unforeseen events. Planning is integral to any successful organisation. Mapping out a path for the future, and anticipating and making provision for change, are fundamental to the business environment. Robbie Burns knew that there is no such thing as a foolproof plan. Perhaps he can be regarded as a true prophet of modern planning.

The planning of a fleet of aircraft for an airline is really no different from any other planning activity. It is fraught with complexity, dilemmas and uncertainty. Building a successful fleet plan requires a blend of engineering and commercial know-how, the ability to predict the future, a good deal of intuition, plus a lot of luck. Good fleet planners, one can almost say, are born, not made.

This book is about fleet planning, and specifically addresses transport aircraft with a passenger carrying capacity of 100 seats or more. Yet we should appreciate that such aircraft are not only acquired by commercial airlines. Leasing companies, governments, financiers and private individuals may also be customers for aircraft.

It is essential that we start with a definition of airline fleet planning.

Indeed, we need several definitions in order to embrace the attitudes of these various customers of aircraft. Let's start with the viewpoint of the airline.

Fleet planning is the process by which an airline acquires and manages appropriate aircraft capacity in order to serve anticipated markets over a variety of defined periods of time with a view to maximising corporate wealth.

However, a leasing company may well adopt the following attitude.

---

1   The best laid plans of mice and men go oft awry, and leave us nothing but grief and pain instead of promised joy.

The acquisition of a fleet of aircraft is a means to provide a portfolio of opportunity to profitably rent to organisations requiring flexible solutions to their aircraft capacity needs.

Already we have two very different objectives of customers seeking the same product. We could go on and expand these definitions even further. For example, there are very different types of airline. Scheduled, leisure, low-cost or cargo airlines are all driven by a multitude of objectives according to their particular markets and can consequently be expected to adopt different approaches to planning. On the other hand, leasing companies will consider the aircraft to be a vehicle to generate regular rental income and, crucially, enables the company to realise a profit on the eventual sale of the asset.

However, let's concentrate for a moment on our airline-based definition of fleet planning and consider the key words. Firstly, we can consider fleet planning a *process*. We can certainly envisage that a structured review of the airline's market positioning and aspirations must take place. It is essential that any type of structured review take place on a continuous, rather than ad hoc, basis.

Next, we determine that an airline *acquires and manages* its fleet. Acquisition may take the form of outright purchase or rental. The degree to which an airline opts for one or the other could be driven by factors such as financial expediency, spreading risk among different suppliers, the availability of aircraft, or even national pride. It is true to say that fleet planning today has become more than just a matter of acquiring aircraft. The management of that capacity, such as realising an unexpected cash value by selling an asset, is just as much a part of the process.

Our definition goes on to describe aircraft *capacity*. Capacity, in the sense of fleet planning, is a generic term for the size of an aircraft. However, the number of frequencies employed in a market plays a major role in a fleet plan. As we shall see, the understanding of how markets respond to the relationship between aircraft size and frequencies offered is crucial to airline success.

Fleet planning should not be concerned with today's markets alone. *Anticipated markets* must be taken into account too. Planning aircraft size for a market is rather like buying shoes for children. A good fleet plan allows the market to grow into the aircraft just as little feet must grow into shoes. You do not want the shoes to become too small too quickly in a situation of rapid growth. Phasing of capacity to need is very important.

A *variety of defined periods of time* suggests that a fleet plan should be valid for several periods. The essence of this part of the definition is that one particular solution may only be appropriate for a single time horizon. It may be the case that taking a longer-term view would result in a wholly different initial solution. This is because we might invoke the effects of technical obsolescence; for example, or else the imposition of anticipated environmental restrictions. So, a fleet plan should take account of different time periods in order to judge whether the solutions would change. Only then can an optimum choice be made.

Lastly, our definition refers to *maximising corporate wealth*. Although this does sound rather obvious, it is nevertheless the case that many airlines in today's competitive world have pursued market share at the expense of their bottom lines. The temptation to dominate through market share alone is a dangerous phenomenon, and the history of air transport bears testimony to this. It is as well to emphasise that, for the majority of the world's airlines, overall profitability should be the main driver of the fleet decision.

One of the most frustrating things about fleet planning is that no matter how logical, how financially sound, how compelling the case for implementing a particular fleet, the real decision may be driven by purely extraneous political factors. The content of this book is not built around a planning system controlled by unwise or unwelcome intervention by outside agencies, but rather one controlled by cool and old-fashioned logic.

Who Controls Fleet Planning?

There is an old joke that in the early years of airlines, it was the pilots who bought aircraft. Then we went through a phase in which fleet planning was driven by engineers. After that, the marketing people got their hands on the process until finally the financiers took control. It is certainly true that we went through a phase in the 1980s when the fleet planning departments of many of the world's airlines were composed largely of engineers. They made solid engineering assessments of the products on offer, calculated take-off and *en-route* aircraft performance, then evaluated airport compatibility, systems reliability and maintenance requirements. Cool and old-fashioned logic, as mentioned above, would then lead to a rational, and technically justified, solution.

Life is not, of course, that simple any more. Firstly, the impact of deregulation and increased competition has changed airline priorities. An airline's chief priority today is not to supply air transport as a service, or act as a representative of the state, but to act as a commercial business. Of course, there is one element of air transport, safety, which will always be rightly regarded as *the* top priority. Beyond this, today's airline is now a commercially, rather than technically-orientated business. The need to generate profit to stay in business has meant that airlines have been forced to become more nimble in order to adapt to rapidly changing market conditions. So the focus has shifted away from the engineer to other players: the commercial teams, who determine route structures, fare policies and brands; the financial controllers, who determine investment levels, sources of funds and whether any money can be spent at all.

These changes have consigned the engineer to a supporting role in the fleet planning decision process. Luckily, he is not museum-bound, for his judgement is still fundamental.

In the same way as airline competition encouraged an evolution in management priorities, we have seen changes in the way airlines structure their fleet planning organisations. In pre-deregulation days it was not uncommon for most major airlines to include substantial fleet planning departments in their organisations. New business models throw much more emphasis on supporting only the core activity. Anything else can be outsourced. It is not entirely surprising that classical airline fleet planning departments have been stripped down to a shadow of their former selves, to be replaced by a new breed of asset managers, with a much broader spectrum of activity.

Fleet planning is fast being supplanted by fleet management. As a consequence, we should now separate the acquisition and management activities. Owing to the greater complexities of aircraft evaluation and acquisition, involving a variety of functional departments, today's fleet planner has become a project manager, rather like the conductor of a symphony orchestra. He or she is selecting the players, ensuring that everyone plays in the same key and remains in tune. The conductor allows each player an appropriate voice at the right time, and makes sure that all those involved finish at the right time.

Where Should the Fleet Planner Fit in the Organisation?

There are numerous answers to this question, of course, depending upon the type of airline, corporate structure and objectives, cultural background, and even size.

Scheduled and cargo airlines favour a more formalised fleet planning activity, whereas low-cost carriers, charter airlines or leasing companies clearly see their fleets as vehicles to achieve different goals.

The tendency of fleet planning to become more market-orientated will also influence the decision. Also, as airlines now tend to place more emphasis on the financial, as opposed to the technical, side of the evaluation, one can expect fleet planners to be more closely allied to the financial side of the business.

Fleet management tactics, such as aircraft trading, are also pulling the planning process in a new direction. There is a strong argument to keep the pure asset management aspects of fleet planning separate from the evaluation side. Asset managers will wish to identify the optimum time to sell an aircraft, in order to realise its market value, or else lease-out capacity. However, the traditional evaluation team is more concerned with market fit and technical adaptability. There is certainly a blurring of the edges between fleet management and fleet planning, as the two processes must work in concert. The organisation should take account of this.

A large airline is unlikely to approach its fleet planning in anywhere like the same degree of detail as a small airline running a fleet of, say, a dozen aircraft. The orders of hundreds of aircraft placed by the likes of easyJet and Ryanair, for example, are not made on the basis of a precise knowledge of which routes shall be operated. Similarly, the emerging mega-carriers in the Middle East, such as Emirates and Qatar Airways, are adept at placing significant orders in order to secure delivery positions and commercial conditions in the face of very obvious growth opportunities. However, when these orders are placed it is far too soon to have a clear idea as to where the capacity will be deployed in any detail.

On the other hand, a small carrier operating a handful of routes and aircraft is clearly more likely to have a picture of its longer term capacity needs as the scope of the problem is much reduced. The analytical challenge in this case would be very different, with more focus on real operational needs and specific route forecasts.

Leasing companies often acquire capacity on a purely speculative basis, with no customer in mind at the moment that an order is placed.

The levels of risk in above cases may vary considerably and the fleet planning needs will be driven by different factors.

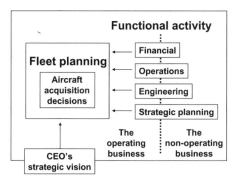

Figure 1.1   Nesting of Fleet Planning

Reporting Lines

Imagine that you are in the shoes of the aircraft manufacturer, or perhaps you are an interested party within an airline, and your task is to establish who is the person most likely to carry the overall responsibility for fleet planning. You might confidently knock on the door of the Head of Strategic Planning. Certainly, you would be welcomed with opened arms and be told that you had come to the right place. You would be informed that the Strategic Planning directorate centralises the definition of the airline product, brand and market segmentation as well as forecasting activities. As a consequence, the definition of the cabin of the aircraft is a vital ingredient in market success and the fleet decision could involve radically reshaping overall company strategy. Clearly, you would be told, route selection and aircraft selection are inextricably linked and it is obvious that fleet planning should be driven by this key department.

So far, so good, except that you are aware that the office next to the Head of Strategic Planning belongs to the Head of Flight Operations and you therefore decide to pay him a courtesy call to see if he has any interest in being involved in fleet planning.

Upon entering his office you find yourself immediately bombarded with arguments that fleet planning should be in the hands of the Operations directorate. It will be pointed out that the schedule is at the very heart of the airline and the fleet decision would impact critical and labour-intensive areas such as crew resources and planning. Indeed, the entire operation can only run smoothly if the fleet is properly planned by the Head of Flight Operations, who will also define the ground rules for aircraft comparisons, assess flight characteristics, check the credibility of performance estimates and guarantees, assess training needs, crewing requirements and avionics needs.

As you exit the office of the Head of Flight Operations you recall that, in days gone by, the Head of Engineering used to have a role in fleet planning and decide to check whether this might still be the case. Now your troubles really start. The engineering chief essentially gives you a rocket and explains that the maintenance of the fleet is pivotal to the success of the airline and only technical experts can give an unambiguous and neutral recommendation as to which types should be operated. Furthermore, there are safety considerations that need to be assessed and the Head of Engineering is uniquely qualified to manage these. Also, he must select vendors, detail the aircraft specification, estimate maintenance and spares requirements, determine product support packages and ground equipment needs.

You now sheepishly leave the office of the Head of Engineering, thoroughly confused by what is obviously a series of compelling arguments. Your eye is caught by a door marked, Chief Financial Officer (CFO). Instinctively you draw back, but pluck up courage to poke your head around the door and ask the same innocent question, 'Are you involved in the fleet planning process?' The CFO adopts that air of superiority found lurking in many CFOs throughout the world. He says, simply but convincingly, that he is the one who authorises all expenditure and, irrespective of anyone else's point of view, he shall have the final word. The CFO will also point out that the magnitude of aircraft investment is such that the financial viability of the whole enterprise might be at stake. Only the CFO can validate the economic inputs, evaluate various financial offers and ensure that the portfolio is managed in such a way that potential tax advantages are realised. He will tell you that the only way a fleet planning decision should be made is according to whether the investment makes overall sense in terms of net present value, and that this element is more important than anything dreamed up by the marketing team, pilots or engineers in the airline. The CFO would need to perform a financial risk analysis to determine capital investment needs, taxation and depreciation implications, and to seek out any other aircraft trading opportunities. He will also explain that even in the face of a compelling argument to change the fleet type, external financial conditions may mean that it is the wrong time to invest.

Perhaps some of the above rings true? In fact, all of the arguments put forward by the various heads of functions are valid. Our little scenario simply proves that fleet planning is a multi-functional activity that must be overseen by a principal director, who must organise and coordinate the inputs and arguments of a variety of players.

However, the complications deepen when we consider that there are sometimes non-operational implications of a fleet change that go beyond the direct effects of an aircraft on the operation itself. For example, when Air New Zealand was assessing the acquisition of a fleet of A320 aircraft, they had to consider the implication on their profitable third-party maintenance services for small operators of 737 aircraft in the Pacific. Put simply, they had to evaluate the consequences of losing 737 knowledge in the event of a fleet change, against the additional cost of maintaining their 737 knowledge, but needing to acquire A320 maintenance expertise to support their own fleet. As another example, Singapore Airlines has been known to make more non-operating profit than operating profit thanks to their skill in managing their fleet financial portfolio.

To sort out the various arguments and different opinions, there should be a central and overriding decision-maker. This person is often the Chief Executive Officer (CEO), or Managing Director. He, or she, is able to ensure that all voices are heard, that there is a hierarchy of intervention and that the fleet decision fits into the airline needs from an overall strategic standpoint. For example, it might be considered appropriate that, notwithstanding the attributes of any particular aircraft, that the airline should pursue a dual supplier strategy in order to ensure that aircraft manufacturers do not become complacent, or else dictatorial in terms of pricing and

Figure 1.2   British Airways Organisation

innovation. The decision by Singapore Airlines to select both the A350XWB and the 787, which are aircraft competing against each other, is an example. In this case, the airline is sufficiently large to be able to successfully integrate two different types without operational or economic penalty.

In a small airline, the CEO would very likely make the decision based upon his own analysis and perceptions, rather than those of a team reporting to him. A 'one-man-show' raises the stakes for the manufacturer by a large margin.

Let's take a brief look at how British Airways organises itself, as an example. Essentially there are three main persons specifically charged with the fleet planning activity. This team, along with the Network Planning and Revenue Management teams, reports to the Commercial Planning directorate, who in turn reports to the CEO. Network Planning is integral to the fleet planning process, incidentally. Responsibilities for aircraft purchasing, aircraft trading and financial analysis fall under the Finance organisation, naturally. Operations Planning takes care of operations control issues, ground customer services, catering, cabin services and crew planning. Engineering manages fleet planning from the perspective of aircraft and engine maintenance, cabin interiors and performance. Flight Operations offers yet another, independent, voice to the melting pot.

Lastly, input concerning legal, public relations, economics, safety, the environment, airports policy, alliances are funnelled in to the analysis when required.

What is the Role of the Manufacturer?

Airline organisational changes have had an impact in the way suppliers interact with their customers. Manufacturers now play a more central role in many fleet decisions, and mirror airline organisational change, as they have also become equally market-orientated. The effect of all this is that manufacturers now shoulder more of the analytical work, and address fleet planning from a far more commercial perspective than in the past.

A partnership between the manufacturer and airline is essential. It goes without saying that the manufacturer has a vested interest in selling his product. For this reason alone we might imagine that the old adage *caveat emptor*, or buyer beware, should apply. However, a professional approach should ensure that both sides work in an atmosphere of mutual respect. No manufacturer would impose his product in such a way that the purchaser's business would knowingly be at risk. That would be disastrous for all concerned. Obviously, the manufacturer will focus heavily on attributes of his product that he perceives as being superior to those of the competitor, and consequently downplay anything that might appear detrimental to his case. Airline fleet planners will therefore very often be presented with sets of argumentation that are contradictory. Frankly, all of this can usually be resolved by careful definition of the assumptions under which analysis is performed. This is a delicate issue and we shall return to it in Chapter 2. Suffice it to say that the airline will often need to balance opposing viewpoints.

The degree to which the airline and manufacturer work together is a function of a number of factors. Firstly, small or start-up airlines will lack the expertise, experience and access to data in order to conduct a comprehensive analysis. In this case the manufacturer clearly has a stronger influence. Secondly, if an airline is already operating a product of a particular manufacturer, then one can argue that the manufacturer has a head-start in further developing his influence.

Why Manufacturers and Airlines have Different Timing Perspectives

Ask anyone working for an airframe manufacturer for a definition of 'long-term', and you will be told, 'anything from ten to twenty years'. Ask the same question of a scheduled airline planner, and the answer is likely to be, 'to the end of the next timetable period; let's call it eighteen months'. The reasons for such a diversity of opinion are clear. An aircraft manufacturer is obliged to consider very long time frames owing to the nature of his product. The conception, design, testing and production of an aircraft developed from a clean sheet of paper can occupy a decade or more. However, the emergence of the family concept has considerably shrunk the airframe manufacturer's development cycle.

The Life of an Aircraft

As an example of how an aircraft develops, let's take a look at the Airbus A320. The idea for such an aircraft emerged at the end of the 1970s, when it became clear that a substantial market would appear during the next decade when first-generation DC-9, 727 and 737 aircraft were due to be replaced. Fundamental decisions had to be taken about the level of technology to be incorporated into the new 150-seater project, and lengthy consultation with potential customers took place. The project received its industrial launch in 1984 when the configuration was largely frozen and customers were formally sought. At this point, hundreds of suppliers and vendors had committed to the programme and Airbus began to put in place a complex industrial process for detailed design, manufacture and assembly. Another three years passed before the first aircraft flew, and just over another year passed before the aircraft entered into revenue service, in spring 1988. So, almost a decade had evolved from initial conception through to entry into service.

The life of an aircraft is only just beginning at this point. Typically, design margins incorporated into the initial design in order to protect the integrity of the

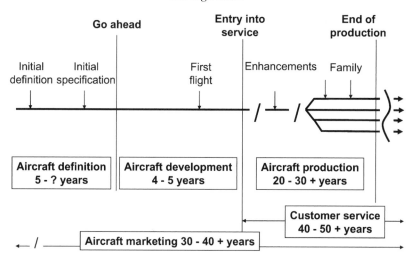

Figure 1.3   Typical Aircraft Programme Life Cycle

airframe structure will be gradually relaxed as experience grows. In other words, design weights will be gradually increased, leading to improved take-off, payload and range performance. Such improvements have the effect of expanding the market of the aircraft. The A320 was introduced at a Maximum Design Take-Off Weight (MDTOW) of 66 tonnes. This gradually expanded through a number of steps up to 77 tonnes.

It is also fairly typical to see stretched and shortened versions introduced with identical technology as market requirements dictate. So, the A320 was joined firstly by the longer-fuselage A321 in 1994, and then by the shorter A319 in 1996, all forming part of the same family. Finally, the last member of the family, the A318, entered into service in 2003 – a fully quarter of a century after the first discussions took place for the initial design. Despite this, no one can argue that the A320 family of aircraft is at the end of its commercial life. Indeed, there are plans to incrementally upgrade all the aircraft in the product range, to include new interiors, aerodynamic improvements to the wing-fuselage fairing, as well as engine improvements. A cumulative improvement of around 4% in fuel burn might be expected. If the designers can identify a further double-digit improvement in fuel burn, then the possibility of a more radical product improvement becomes possible. Furthermore, the potential always exists for a 'P2F', or passenger to freighter, programme, thereby further prolonging the life-cycle of the product.

What is clear is that, given the obvious inertia in the process, any serious misjudgement made at the early stages of the product life cycle could prove disastrous for the eventual full development of the potential. Yet despite the inevitable inertia, who could have anticipated in the late 1970s that there would be a totally new breed of airline business model – the low-cost carrier – that would have an appetite for this very aircraft in its more developed form?

Boeing continually upgraded their product lines, to the extent that today's 747-400 bears little technical resemblance to the very first 747-100 of 1969. Similarly, the 737 Next Generation series has also enjoyed remarkable success, becoming a family in its own right, although there is not the same degree of commonality with earlier versions of the 737 as Airbus enjoys with the A320.

Given the magnitude of investment in airframe, as well as engine, development it is little wonder that manufacturers base their product development decisions on a long-term view of the market. Both Airbus and Boeing regularly look 20 years into the future in their *Global Market Forecast* and *Current Market Outlook* respectively.

Although a manufacturer takes a long-term view owing to the very nature of the product, it is also fair to say that the manufacturer must anticipate the needs of his customer and his customer's market. Thus, the factors which drive an airline to gaze a long way into the future are of direct relevance to the manufacturer as well.

In complete contrast with the manufacturer, a scheduled airline sees its future in more immediate terms. Unsurprisingly, an airline's fortunes are linked to what it can achieve in the immediate future and peering 20 years into the future is not exactly a priority. Nevertheless, long-term planning is essential, as ignoring future shifts in market needs may compromise strategic direction. One of the manufacturer's first tasks in undertaking a fleet planning exercise with an airline is to establish his credibility in proposing what may initially be viewed as an unreasonably long-term solution.

Why the Fleet is a Long-Term Issue

Even the most hardened scheduling professional will appreciate the need to make fleet provisions beyond the next timetable period, but the reasons for the airline taking a long-term view do not coincide with those of the manufacturer. Let's consider what drives an airline to consider the fleet as a long-term tool.

Firstly, future capacity needs ought to be estimated so that expected market demand can be accommodated. Indeed, the whole basis of the fleet plan should be that enough capacity be on hand to cope with forecast traffic.

Secondly, it is increasingly the case that fleet planning is becoming synonymous with the overall business plan. This is because the aircraft is seen as a tool that can influence the behaviour of the market. Some will disagree with this, arguing that the aircraft is merely a commodity. Indeed, it can be argued that it is the way in which an aircraft is configured that creates the brand, rather than the aircraft itself. It is still true, however, that the aircraft design has to permit an airline to communicate its brand. If we accept that fleet and business planning are essentially integrated, then it follows that any company that is based around five or 10-year business plans must make long-term decisions about the fleet composition. Any kind of future infrastructure need may impinge upon the fleet decision.

Thirdly, an airline is always looking for flexibility to adjust capacity and frequencies to rapidly evolving market conditions. The fleet is one of many tools to achieve this. The major part of this book is devoted to the correct identification and measurement of market conditions so that an appropriate fleet can be planned.

Fourthly, changing regulations should be monitored as these can certainly impact the fleet plan. We can distinguish between technical regulation and economic regulation.

In terms of technical regulation, airspace management changes, such as the evolution of the rules regarding separation of aircraft due to wake vortex, can influence the performance capability and, therefore, productivity of aircraft. Safety being of paramount importance, any new regulation that affects the design and operating weight of an aircraft will also affect its productivity. Operating regulations are sometimes different for twin-engine aircraft. Extended Twin Operations (ETOPS) regulations imply operating cost penalties for twins, rather than multi-engine aircraft. We shall examine these issues in more detail in Chapter 6. Needless to say, changes in environmental regulations are important to monitor, especially if particular aircraft types are expected to become outlawed before the end of their useful lives.

In terms of economic regulation, the onslaught of liberalisation has resulted in significant changes in fleet structure, with airlines tending to reduce average aircraft size in order to play with frequencies. Changes in bi-lateral restrictions and slot allocation procedures are radically affecting the correct aircraft size for a particular market.

Another reason for an airline to take a long-term view of fleet planning is that future technological development may result in lower operating costs. Gone are the days when airlines acquired aircraft purely to inject new technology into the business for its own sake. Today, prime importance is attached to the ability of new technologies to translate into better efficiency and higher profits. Sometimes new developments offer a completely new size of aircraft, such as the A380 for example, which is having a significant impact on the types of aircraft already operated. The implications of introducing a completely new size of aircraft are far-reaching. Elements such as the asset values of competing types, infrastructure access, airspace use, route and network economics are all affected to a great degree by a new size of aircraft. The introduction of the 747 in 1969 had an overnight and dramatic effect on airline operating economics as airlines tussled with a sudden explosion of capacity. Also, Boeing's innovative but ultimately misjudged Sonic Cruiser concept in 2001 looked like it would reshape airline networks, although having an unpredictable impact on the values on conventional aircraft it would have replaced.

Resolving Conflicts of Time Perspectives in Fleet Planning

We have seen that the aircraft manufacturer and airline both need to take a long-term view of fleet planning, though for different reasons. To conclude this perspective of timing, let's assess the relationship between short and long term, and market needs balanced with the provision of resource.

Essentially, the planning process is one of balancing the degree of adaptability of a company's resources against the predictability of the market and the environment. In the short term an airline has relatively little flexibility in adapting its resources. The airline business has traditionally been seen as capital-intensive, requiring significant investment in aircraft and associated facilities. Adaptation becomes easier as time goes on. Therefore, in the short-term the best that can be achieved is an improvement in the aircraft assignment to particular flights, or else a change to the timetable. In the medium-term we might envisage an alteration to the route structure and market served. Only in the longer term can we plan for a complete fleet roll-over, for example,

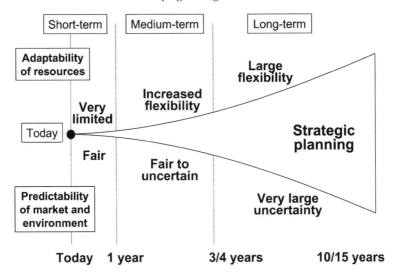

Figure 1.4   Conflicts in Timing

or else change in the business model itself. Conversely, the market and environment in which an airline operates is only truly predictable in the relatively short term, and uncertainty expands with time. Strategic planning is fundamental in closing the gap between growing flexibility of resources and growing uncertainty of the market.

Adapting to Timing Differences

However, the industry has found its own solutions to this perennial challenge. Firstly, airlines have become much more nimble in matching their resources to market changes. The speed of response to a new market condition is influenced by a variety of factors such as airline size, management style and the political and economic climate. Secondly, the supply of aircraft capacity to the airlines has become more flexible. Manufacturers have striven to reduce the time from placing an order to delivery. It is not uncommon for the lead-time between the placing of an aircraft order and delivery to be as little as 9 months – at least for a single-aisle aircraft where customisation issues are not as challenging as they are for long-haul twin-aisle aircraft. Lead-time for delivery depends upon whether an order is for a brand-new type or merely building on an existing fleet. It is quite common, at least for single-aisle aircraft, for an airline to select the generic type within the family and make a late decision as to what the delivered aircraft type should be.

Airlines have also become adept at managing their order books by committing to a balanced mix of both firmly ordered and optioned aircraft. A firmly-ordered aircraft is contractually committed, whereas an optioned aircraft is subject to a future and separate negotiation. The purchaser will certainly have to pay for the privilege of having some delivery slots for optioned aircraft. However, finding the appropriate balance between the number of firm and option aircraft can be a delicate issue and the manufacturer can have an ambivalent attitude towards the issue. An optioned

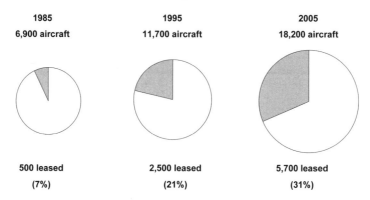

1985	1995	2005
6,900 aircraft	11,700 aircraft	18,200 aircraft
500 leased	2,500 leased	5,700 leased
(7%)	(21%)	(31%)

Figure 1.5   The Impact of Operating Leasing
Source: Airbus

aircraft is effectively a liability for the manufacturer as he is still unsure whether the airline will exercise the option in the future. Indeed, the track record seems to be relatively poor, with perhaps only half of optioned aircraft finding their way eventually into the order book. However, the inclusion of optioned aircraft does help boost the manufacturer's potential order book and give the buyer some leverage in the overall negotiation. There is no particular rule for the correct balance between firm and option aircraft. A 50:50 split is very common for rapidly growing airlines, whereas smaller, more stable operators would perhaps be content with a minimum of options. One reason an airline might wish to restrict the number of option aircraft in an order is to dampen the future investment requirements. Sometimes it is not in the interests of the airline to spook their shareholders concerning the degree of investment capital required in the future.

The leasing companies provide another safety valve in controlling the speed at which capacity comes into the market and, crucially, in providing capacity at relatively short notice. Both Boeing and Airbus have seen their customer base considerably expanded thanks to the creation of new markets by the lessors. Operating lessors see aircraft as a financial vehicle rather than a transport vehicle and are just as influenced in their buying strategies by the economic cycle as the demand for air travel. Their strategies are to rapidly develop portfolios of young and flexible aircraft and offer aircraft for delivery with as little as three or four months' lead time. It is now quite typical for the major manufacturers to allocate almost one third of their capacity for the lessors. Neither Boeing nor Airbus would wish to cede total market power to the lessors, but the role of the lessors is nonetheless vital. Indeed, in 2005, 16% of all aircraft orders for aircraft of more than 100 seats were made by the leasing companies.

Coping with the Cycle

Ask any forecaster for a view of the future and the odds are that you will be shown a beautifully smooth curve, trending optimistically upwards. Forecasters are, of course, just as aware of the effects of cyclic variations in demand as anyone else, but

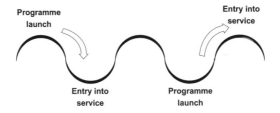

The manufacturer's dilemma

Figure 1.6   Boom and Bust Cycles – 1

it is just not possible to predict with any certainty the precise timing of an economic downturn, currency crisis or war. After the terrorist attacks of 11 September 2001 demand forecasts were thrown into disarray. However, once traffic had recovered to the pre-crisis levels forecasters felt comfortable in maintaining their view that the rate of growth would be maintained at the same levels as before the crisis. Whilst it will always be impossible to predict events as dramatic as those of 11 September 2001, we should be better equipped to predict the consequences of them.

Forecasts give underlying trends based on a set of parameters and explanatory variables. However, it is useful to review how market dynamics have been affected by external parameters with history as our guide.

Although we cannot eliminate the effect of the economic cycle, we would do well to heed its effects on both the manufacturer and airline. Unsurprisingly, airlines tend to order aircraft, and manufacturers tend to launch new programmes, when times are good. Equally unsurprisingly, first deliveries of new aircraft types have a tendency to occur during an economic downturn. Far from being a result of bad planning, this is purely a function of the rather longer lead times involved in aircraft production compared to the volatility of world, or regional, economies.

It takes a great deal of courage and foresight to launch a brand new aircraft type, when profits and business confidence are at a low ebb. Clearly, the launch of a new aircraft programme involves huge investment. Both the 787 and A350XWB programmes are consuming investment in the region of $10bn. The A380 development costs were much more. Maximum confidence in the potential for new aircraft is unlikely to be found at the bottom of the business cycle. It is usually the case that the massive commitment needed for a new programme will only be made at the top of the cycle. Yet, once the button has been pushed, a manufacturer is suddenly at risk of seeing his product launched when business conditions are weakened. Early deliveries of a new type are a vulnerable moment in the life of a new product and it is essential to maintain a successful ramp-up of production in this phase of the development. The A380 was fortuitously launched in December 2000, before the airline crisis of that decade began to bite hard. It is very typically the case that after an initial flurry of orders, things would calm down during the production of the first aircraft. The order book would usually remain stalled for a period for several reasons. Firstly, early delivery positions might be sold out for some time and there would be nothing to be gained by placing an order. Secondly, manufacturer's launch conditions would very likely have ceased at a certain moment. It is logical that airlines that have failed

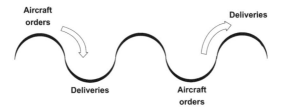

The airline's dilemma

Figure 1.7   Boom and Bust Cycles – 2

to secure launch pricing would prefer to wait until the aircraft has been delivered to first customers so that a better appraisal can take place. Thirdly, the timing of the economic cycle may play a role, as it is imperative that orders take place against a backdrop of economic and market confidence.

There is a remarkably close relationship between airline profits and orders, and an equally remarkable lack of synchronisation between profits and aircraft deliveries. Indeed, we may identify a very similar mismatch between economic cycles and airline decisions as we observed between economic cycles and manufacturer decisions.

The airline industry seems to find itself the victim of this problem on a regular basis. In the late 1970s (Figure 1.8) we can observe an increase in aircraft orders as a result of two elements: anticipated replacement of single-aisle aircraft ordered in the previous decade; and the effects of the newly-deregulated environment in the United States. When these aircraft were delivered at the turn of the decade the industry found itself victim of record-breaking fuel prices and economic downturn. As a consequence, airlines were losing money for the first time in history. Recovery was painful as airlines were drawn into damaging price wars in order to fill up capacity.

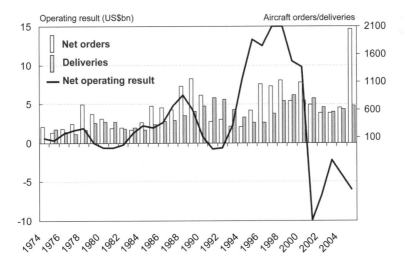

Figure 1.8   Desynchronised Cycles
Source: Airclaims, airline reports, IATA

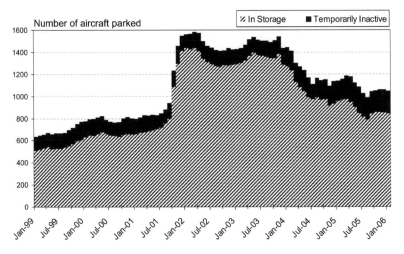

Figure 1.9   Stored Aircraft
Source: Airbus

When the next economic downturn came round at the beginning of the 1990s, we can observe a similar situation. Only this time the leasing companies had exacerbated the situation by ordering large numbers of aircraft on a largely speculative basis. The crisis was much deeper this time and the number of parked aircraft reached record levels. Once again the airline business showed its resilience by recovering quickly, although this was a painful process as there had been significant over-capacity. The most recent downturn of the 2000s followed another period of large orders for aircraft. Yet something dramatic occurred in 2005. The industry was still globally making losses (although most were concentrated in the United States) but orders reached an all-time high of 2140 aircraft (in the category of 100 seats or more). A total of 29% of those orders were attributed to low-cost carriers. The Chinese and Indian markets had burst onto the scene to such an extent that over 30% of all firm orders in 2005 were destined for those two markets alone.

Although the natural economic cycle will always be with us, disparities between deliveries and profits are likely to be more in tune in the future as the wave of replacement cycles dampens down and manufacturers and buyers more closely anticipate the release of capacity into the market.

The Problem of Stored Aircraft

Airlines love to blame the manufacturers for over-production. In reality, the manufacturers never plan to build aircraft on a purely speculative basis, and always have customers lined-up and willing to pay for aircraft coming off the production line. However, owing to the unpredictable nature of demand, the cyclic and highly competitive nature of the business, there are times when a finding a home for an aircraft proves impossible. Another reason for aircraft finding their way into store is due to technical or economic obsolescence, in which case such aircraft would be unlikely to find their way back into action.

At the time of the first Gulf War, British Airways deferred the delivery of $1bn worth of aircraft. During the SARS crisis of 2003, Cathay Pacific grounded up to 43% of their entire capacity. After 11 September 2001 many airlines simply took the opportunity of grounding the more venerable members of their fleets.

Size Matters

Things are getting bigger. Airlines have consolidated to become giant and formidable competitors, with the promise of even greater dominance as the new breed of airline alliances emerges. Airport hubs have grown to enormous proportions as airlines restructure their networks, particularly in the Middle East. More recently, we have seen the onslaught of low-cost carriers, who enjoy huge rates of growth and order gigantic numbers of aircraft. The A380 is now in service with significant potential to reshape the business. Airlines are continuing to consolidate their activities.

What of the aircraft themselves? In some markets we can observe a downward trend in aircraft size but, as the development of frequencies must ultimately be limited, average aircraft size is set to rise over the longer term.

The airline business is undoubtedly consolidating in many ways. The number of individual deals may not grow at the same pace as the world fleet. This may be to the benefit of the buyers, who believe that they can extract discounts from the manufacturers. Certainly, there are economies of scale to be generated through spares investment, maintenance and operating practice, crew costs and a host of other elements that contribute to efficiencies of fleet operation. It must also be said that size in itself does not guarantee economy. Indeed, as airline size grows, so does bureaucracy, decision-making, and overall inertia. Let's reflect on three of these key issues: strategic alliances, hub-and-spoke operation and low-cost carrier fleet planning strategies.

Strategic Alliances and Fleet Planning

Since the late 1990s the airlines have become embroiled in alliance frenzy. The reasons for this are two-fold. Firstly, alliances are viewed as an efficient means of infiltrating new markets by means of a partner's network. Secondly, alliances hold the promise of cost reduction through economies of scale. Whether alliances are anti-competitive or truly beneficial to the passenger is a broad issue and open to considerable debate depending on which side of the fence one sits. The big question in fleet planning is the degree to which alliances might affect aircraft selection decisions and, crucially, aircraft acquisition. From the airline point of view there is clearly an attraction in joint sourcing of aircraft. However, there are numerous practical objections.

Impediments to joint purchasing Firstly, it is highly unlikely that just because a group of airlines form an alliance that their fleet planning needs are going to coincide. Each and every airline is unique in terms of its strategic approach, its market, goals, and overall structure. This suggests that any form of joint approach in aircraft acquisition would only be possible if one or more partners would be willing to compromise. How this might be achieved is dependent upon the degree of domination of the main parties to the alliance.

Secondly, manufacturers will resist transferring contractual terms dimensioned for a major airline to a much smaller operator within an alliance. Each contract is individually tailored and pricing levels are determined according to conditions that are unique to each case. Manufacturers are not willing to see their pricing policies eroded, if not destroyed altogether. Furthermore, both manufacturers and individual airlines are bound by confidentiality clauses and the sharing of tailor-made deals would be frowned upon. Pricing of aircraft deals is under enough pressure, without the added burden of alliance-driven pricing as well. Indeed, even strong airlines with bargaining power would accept that their discounts are partly possible because other smaller airlines are paying higher rates.

Thirdly, airlines tend to be fastidious in their aircraft configuration requirements. Sometimes this is due to brand building, sometimes it is due to their airworthiness authority requirements, and sometimes it is due to union issues. Nevertheless, the manufacturer always has scope to reduce production costs by means of simplified aircraft specifications and configurations. For this to happen, alliance partners would need to compromise in terms of their requirements for flight deck design, avionics, cabin interior specification, and a host of other customisation items. As an example, Lufthansa require that certain cockpit switches operate in a particular direction. A seemingly small issue has implications in terms of design, production and operation that are specific to this airline. Experience to date suggests that airlines are not ready to compromise on their aircraft definitions for the benefit of their alliances.

On the other hand, if alliance partners *can* agree on common specifications that would have a measured effect on the manufacturers' costs of customisation and production, then opportunities might exist for all parties to gain from a bulk deal. However, entrenched views, union and regulatory requirements, and individual market needs, all conspire to make joint specification difficult to achieve in reality. The Star Alliance has made progress in this area by agreeing on a common specification for a regional jet. One further benefit that can be realised is the exchange of information between alliance partners concerning aircraft design, performance and operations.

Successful joint purchases   Despite the apparent difficulties it is worth remembering that joint purchases within an alliance have already occurred. The now defunct Swissair and Sabena once jointly purchased A330s, for example. Also, it is by no means necessary for airlines to be united in an alliance for a joint purchase to succeed. A significant joint commitment for Airbus aircraft were made by LAN Airlines, TACA and TAM. It was all a question of agreeing on a common specification and being flexible in terms of delivery scheduling. One can also point to other examples of fleet planning decisions being undertaken where equity stakes are held. For example, Emirates was involved in Sri Lankan Airlines' A330 acquisition.

Opportunities for co-operation   The importance of global strategic alliances does mean that fleet planning needs to take account of potential synergies that could arise through common aircraft usage and overall fleet management. In reality it is

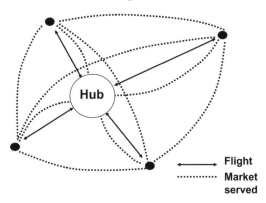

Figure 1.10 The Power of Connectivity

more likely that cooperation be limited to purely technical issues as the issue of cabin design is more personal to each airline, irrespective of any potential alliance synergies. Thus, Qantas and British Airways continue to pursue very different cabin strategies, even though they are partners in Oneworld.

It is not excluded that future aircraft designs could become a catalyst for a new form of acquisition structure for the major alliances. This might be possible if a new design could be regarded as a requirement for *all* alliance partner members at the same time. If needs truly coincide, and if the manufacturer clearly sees a path through which economies of production are present due to airline standardisation agreements, then some form of joint purchase cannot be excluded.

The strategic alliances seem to be entering into a new phase. On the one hand they have finally cemented their identities, but on the other they are more and more seen as being transitory devices that have failed to deliver the full benefits once envisaged. The withdrawal of Aer Lingus from Oneworld is a good example of an airline that has shifted its focus away from being a major international player toward that of dealing with low-cost competition.

Hub-and-Spoke Networks

Another aspect of size and concentration is the development of hub-and-spoke networks. All over the world airlines have created hubs in order to magnify the number of connections they offer. The theory of the hub is that a set of flights converges on a hub airport within a specified wave, so that connections can be made within a certain minimum connect time, and then flights disperse along the spokes to their destinations.

Hubs create tremendous connection opportunities due to the mathematical effect of matching large numbers of arrivals and departures. Thus, a group of as little as four aircraft arrivals, each creating four aircraft departures, creates 10 city-pair connections, as shown in Figure 1.10.

Two hubs, each with four arriving and departing flights can connect 45 markets, as seen in Figure 1.11. Multiplying the number of arrivals and departures within a connecting wave can quickly create thousands of potential market connections.

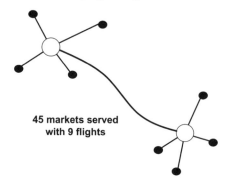

Figure 1.11 Double Hubbing

Hubs tend by their very nature to be monopolistic, with an airport's home carrier becoming highly dominant. They also breed inefficiencies in resource utilisation, as peaks of activity are clustered around the waves. In particular, aircraft utilisation tends to be compromised as the goal of creating connections overrides any economic desire to generate the maximum number of hours' use of the aircraft. Scheduling thus becomes a key issue in determining true fleet economics. As hubs increase in size, so does the probability of delays and missed connections. There is certainly a finite limit to a hub; if operating efficiencies and passenger satisfaction are to be optimised. It would appear that maximum efficiencies occur when around 50–70% of traffic is connecting at a hub. Atlanta and Dallas Fort Worth airports, which are dominated by Delta Air Lines and American Airlines respectively, both achieve connecting passenger numbers within this range.

Hubs are important in the context of fleet planning because the hub strategy, in terms of network design and number of waves per day, will contribute largely to the aircraft size required. Whether an airline uses a small feeder airline to generate traffic for a trunk service, or whether traffic along the spokes is kept at a constant level, will clearly determine the type of aircraft required. An important parameter is the type of hub operation. These can be classified in various ways.

A hub serving a hinterland may receive a series of long-haul flights dispersing traffic to a capture-zone of small cities. In this case, aircraft types using the hub will be a mixture of large and small aircraft. Alternatively, a hub composed of equi-distant spokes serving as a distributor within a region may require aircraft of a more homogeneous size.

Whatever the type of hub it is almost certain that operating economics will deteriorate due to the difficulties in optimising aircraft utilisation. All but one of the US majors operates hubs. Southwest Airlines' linear route network allows them to achieve very rapid turnaround times.

Although hubs have become essential in enabling airlines to build connections and dominate markets, the problems of hub size have also encouraged airlines to grow their business through hub by-pass strategies. The emergence of regional jets with enough range to link a substantial number of small city-pairs has enabled airlines to exploit niche markets and has enabled passengers to enjoy faster point-to-point journey times, without the inconvenience of going through the hub. Crucial to a by-pass strategy is the existence of a market large enough to justify any service at all.

It is worth noting that the two major airframe manufacturers have divergent views on the degree to which airlines will continue to develop hub strategies in the longer term. Boeing firmly believes that airlines will wish to concentrate on what it terms 'point to point' flying, which suggests that airlines will need to focus on relatively smaller aircraft to link a larger number of direct markets. However, Airbus believes that hubs will remain an important component of network design, which suggests a need for larger aircraft that offer the capacities needed to efficiently connect major centres of population. Airbus also insists that some fragmentation will take place in parallel. A cynic would argue that these views are designed to support the product strategies of the two suppliers. The truth of the matter is that there is more than a grain of truth in both approaches and the situation is certainly not black or white. Although both manufacturers' forecasting teams are in fairly broad agreement concerning the overall rate of traffic growth, their divergent views on the types of aircraft needed to support this growth means that each supplier is predicting very different numbers of aircraft in the long term.

The Onslaught of Low-Cost Carriers

There is no question that low-cost carriers are changing the airline business in an irrevocable way. They have placed orders for significant numbers of aircraft and are clearly a dominant feature of the order books of both manufacturers. What concerns us is how their fleet acquisition decisions are made.

In the early days of low-cost carriers there was a general feeling that the 737 was the aircraft of choice. This was principally due to the fact that Southwest Airlines decided to focus on that aircraft from their inception in 1971. An erroneous view had developed that low-cost carriers could only function efficiently if they operated a single aircraft type. These views have now altered, principally from the moment when easyJet decided to order the A319. It is now clear that both manufacturers' products are well-placed in this business sector, and it is also clear that a dual-type fleet strategy can be appropriate for these airlines. We shall address low-cost carriers' attitude to fleet planning in more detail in several Chapters of this book.

Anatomy of a Campaign

It is not the purpose of this book to analyse the full complexity of a campaign to sell an aircraft to an airline. We shall focus mostly on the analytical process to recommend a particular aircraft type, in the overall context of building a fleet plan. However, we do need to spend a moment reflecting on where this analytical process sits alongside the many other relationships that exist between airline and airframe manufacturer.

Typically, there is always some sort of relationship between the airframe manufacturer and the prospective customer. This may exist on a rather informal basis, during which the supplier will simply provide the airline with updates on his product and try to extract ideas from the airline on his need. One might call this phase, the 'watching brief'. The manufacturer should beware of providing unsolicited information in the form of study work that might be misconstrued as being a response to an airline request. Airlines can be quite sensitive to this as they would not wish to convey the wrong impression to any prospective supplier.

Depending on the situation, there may be a Request for Information (RFI), which is a more formal way of channelling a prospective purchaser's questions in a structured manner.

Formality takes hold when a key document is prepared and issued to the manufacturer. This is the Request for Proposal (RFP). We shall return to both the RFI and RFP in more detail in Chapter 4.

The RFP marks the start of the campaign proper, where milestones need to be set on both sides and where priorities need to be identified. Various alternatives for study need to be identified and information and analysis performed to whatever level of detail appropriate. The provision of the assumptions under which the network, market and aircraft should be analysed is critical to the outcome. It would not be uncommon for the airline to provide similar assumptions to both competing manufacturers.

Some airlines, notably Emirates, prefer to orientate their study assumptions differently so that each manufacturer can simulate their aircraft in the best possible conditions, rather than having to compromise to meet specific airline conditions. The manufacturers are thus free to present their cabins with the number of seats abreast optimised to the fuselage width.

Large airlines prefer to limit the manufacturer input to the very basics, such as fuel burn, aircraft pricing and maintenance cost, and then perform their own internal analysis. Smaller airlines will clearly lack the expertise and modelling skills to undertake sophisticated study and, in this case, the manufacturer's role would be deeper. It is often a good idea for the airline to cross-check the often conflicting inputs from the manufacturers with an internal analysis, or else engage an independent consultant to provide an unbiased view.

During a typical campaign the fleet planner should coordinate all contacts and presentations made by the manufacturer. These may take place at a number of levels. Working level contacts are vital in order to ensure that assumptions are being correctly adhered to and that all relevant areas are being addressed. Senior management presentations are important in order to inform decision-makers of the evolving situation. Similar presentations might be necessary for the airline owners, or the Board. It is typical that the further one progresses up the chain of command, the less detail is necessary!

Once the aircraft evaluation has been completed to the satisfaction of all parties a proposal is made, comprising the type and number of aircraft, delivery dates and conditions and pricing. It might be the case that the proposal is made before the completion of all the analysis. There follows a period of internal reflection on the offer, usually time-bound, before a decision is made. In the event of a positive decision a Letter of Intent (LOI) or Memorandum of Understanding (MOU) would be signed in order to secure delivery positions. This LOI or MOU would be subject to full contract negotiation which would hopefully result in the signature of the contract, or Purchase Agreement.

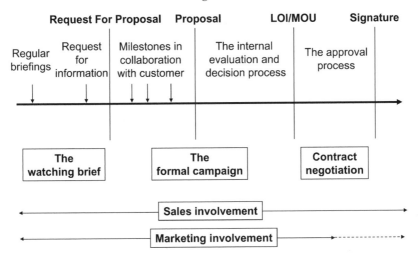

Figure 1.12 Anatomy of a Campaign

There is no such thing as a typical campaign. Each is coloured by the conditions of the airline, degree of competition between the manufacturers, political and economic environment, and hunger of all the parties involved. External political considerations can significantly affect the outcome. This is particularly true in India, for example, where government approval assumes an unusually important part of the overall process, to say the least. Another example concerns countries of Eastern Europe, where states are determined to build a future free of domination – whether from Moscow, Washington, or Brussels. For example, Poland's national carrier, LOT, has frequently found its fleet planning decisions tied to accusations that it has betrayed an ally.

Sometimes, campaigns can stretch over many years. It is no exaggeration to say that people have practically spent their careers working on a single case. At the other end of the scale, an option conversion might be a fairly straightforward procedure. On average, a campaign may typically last from one to two years. Anything beyond this would mean that market conditions would begin to change, perhaps invoking a complete rethink of the situation.

The manufacturer's team would usually comprise the Sales Director and Marketing Analyst or Director, punctuated with specialists according to the specific need, up to the point where a proposal is prepared. These specialists are needed to provide, for example, cabin layouts, weight statements and detailed analysis of spares investment. From the proposal onward, the Contract Negotiator takes a prominent role, aided by specialists involved in guarantees, customer option selection, financing, spares provisioning, training, and other technical support areas appropriate to the case. It is also important for the airline to manage its own contacts with the engine manufacturers and vendors of equipment in order that selections can be made.

Beyond the campaign, the sales and marketing personnel maintain their involvement with the fleet planning representatives of the airline in order to provide continuity of relationship with their customer. The manufacturer has an interest in ensuring that deliveries are properly managed and will obviously seek further opportunities to extend the order book. Feedback from the customer in terms of experience of the aircraft in operation is also a vital ingredient in the relationship.

In Summary

We have seen that fleet planning is a highly complex and multi-functional activity that affects the whole airline. However, the degree to which the choice of aircraft directly affects airline brand and image is linked to the business model itself. Thus, low-cost carriers are more concerned with the aircraft as a tool to deliver efficiency, whereas product-orientated airlines see the aircraft as an extension of their brand. Buyers of aircraft also encompass organisations such as leasing companies, who view the aircraft more as a financial than an operating tool.

It is essential to identify at an early stage who are the decision-makers and influencers within an airline. Today's fleet planner is a project manager, whose role is to orchestrate the contributions of many players working in different functional activities.

By its very nature, aircraft design gestation is rather long, and even delivery of an aircraft into a fleet cannot be achieved very quickly. There is always a tendency for a mismatch to exist between the need to rapidly adapt to market conditions and the availability of resources. Reducing lead times is therefore an important objective on the part of both suppliers and operators. Market conditions evolve at an ever-increasing speed, so airlines look for flexible solutions to their capacity needs. The leasing companies play an increasingly important role in this respect.

The industry is going through a period of consolidation, with the continuing expansion of hub-and-spoke systems and alliances. Such developments are significant in that they can colour fleet planning decisions. Low-cost carriers have burst onto the scene and are becoming dominant in more and more markets. This breed of carrier certainly sees the aircraft in a different light than more traditional airlines. Here, the aircraft is merely a commodity, enabling the provision of seats at lowest cost.

Finally, we looked at the anatomy of a campaign and saw that this is far from a mechanical and predictable process. Relationships between the manufacturer and airline must be built up over a long period and maintained well beyond the contract signature.

Before embarking on the aircraft evaluation itself, it is important for any airline to consider carefully how the fleet can contribute to its objectives and goals, and to determine a fleet acquisition strategy. We will now turn to this important phase in the process.

Chapter 2

The Fleet Selection Process

Valuing the Assets

Every airline is, of course, unique in terms of its history, market, philosophy, and fleet requirement. Whilst it is not possible to determine a set of attributes that can be applied to any airline, it is possible to outline a set of principles from which any airline, no matter how big or small, can begin to develop a fleet acquisition strategy.

In the course of setting up the structure of our fleet plan, we shall need to address areas such as: how to strike the correct balance between an overall macro approach and a bottom-up micro approach; and how to determine and measure a set of priorities to elements that make up the plan.

A good way to start is to determine the value attributes of an airline. These will typically embrace:

- *The network*   Includes the routes, the traffic rights and airport slots.
- *The staff*   Measured by experience, efficiency and motivation.
- *The brand*   Reflecting market perception of the company and expectations.
- *The fleet*   The vehicle through which the product is delivered.

The first three items above contain elements of intangibility, being difficult to value but nevertheless bearing a vital intrinsic worth. The strength of an airline depends upon how these elements help the airline adapt to its market and its competition, how they interact with each other, and how they help the airline achieve its objectives. Even these value attributes are worthless without the existence of air travel demand. How this demand is met is part of a more strategic decision concerning how the company functions in its environment, and how adjustments should be made to keep pace with a constantly evolving marketplace.

In broad terms planning an airline business is concerned with defining overall objectives and goals, assessing the target demand and business environment, generating the supply, and monitoring the achievements. The first of these phases will be dealt with in detail here. The remaining three will be covered throughout the remainder of the book.

Defining Overall Objectives and Goals

Setting corporate objectives is an essential part of any business as it directly concerns how resources are allocated. Corporate objectives also provide clear targets to assess airline performance and implement adjustments if required. Airline objectives are obviously diverse in nature, but they can generally be related to one or more of these categories:

- A Marketing Objective
- A Development Objective
- An Alliance Objective
- An Economic or Financial Objective.

Let's examine each of these four objectives and see how the fleet decision may be influenced in each case.

A Marketing Objective

Different types of airline pursue different marketing objectives, so it is useful to identify distinctions in their goals. For example, scheduled operators carry responsibilities that charter operators do not. The type of aircraft chosen by a scheduled airline so that a timetable can be met, irrespective of demand fluctuations, may well be different from the type chosen on the basis of charter work. The scheduled carrier has the option of addressing a variety of market segments, each with its own characteristics and demand patterns, whereas a charter operator is providing a product that is merely one part of a greater single product – a holiday. Even the scheduled airline is providing something that is one part of another product, but this could amount to anything from a business journey or a vacation. The wide variety of journey purposes can have an impact on fleet planning decisions.

Scheduled and charter operators usually have differing views on how the underfloor space of an aircraft is used. The scheduled airline will almost certainly consider the underfloor space as an opportunity to gain extra revenue from cargo, whereas the charter airline might be operating in markets that preclude this. Also, in the interests of maximising the number of seats on the main deck, it is not excluded that the charter airline transposes 'monuments,' such as lavatories and galleys, to the underfloor. The UK charter operator Airtours is a proponent of this philosophy and has made a net gain of nine additional seats on the main deck of their A330-200s as a result of moving the lavatories to the underfloor. In this way, all but one toilet has been saved in the rear section of the aircraft. Although the underfloor toilets take up two pallet positions, this is of little consequence in the charter market. Apart from the obvious economic advantage, there is an 'amenity gain' as queuing and banging of doors in the passenger cabin have been eliminated. Lufthansa has likewise placed galleys, lavatories and cabin crew rest areas below deck in their A340-600s, as they are able to direct the displaced cargo traffic to dedicated freighters. Boeing has innovative ideas to use the space above the main deck of both 777 and 747 aircraft.

There are other elements in setting market segment goals. If a goal is to serve a large number of markets that connect through a hub, then aircraft selection will need to emphasise the ability to offload and reload baggage easily. Indeed, the greater the need to serve multiple destinations through a hubbing system, the greater becomes the need to develop a fleet strategy based upon maximum commonality. This concept has become enormously important in fleet planning and now underpins the majority of aircraft selection decisions.

Geographical location of the network can also be a determinant in aircraft selection. If a marketing goal is to launch a regional operation in South-East Asia, then an operator must consider that Asians tend to prefer wide-bodied equipment. If a marketing goal is to serve a business market in North America, then the tendency for North Americans to travel in Coach-class for business reasons has to be considered. This is unlike European markets, where Business-class cabins are still preferred within the region. Once more, the type of aircraft selected will adapt differently to these different market conditions.

Lastly, whilst it is truer than ever that today's airlines are purely profit-driven, other marketing objectives can still exist. These include operations driven by market-share goals, public-service orientated airlines, and those existing for national prestige reasons.

A Development Objective

A second airline objective might be to achieve a certain degree of overall size. Size can be interpreted in different ways. It is essential in a competitive market to attain a critical mass so that a target market can recognise the presence of the airline. Critical size targets can be set so that growth can be focused and managed. Start-up airlines have a huge challenge in simply getting recognised by the market, let alone established. Much depends upon whether a start-up intends to exploit a new niche or intends to challenge an incumbent carrier. Many low-cost carriers have established themselves to develop an entirely new market. This is especially true in India where operators such as GoAir and Air Deccan are developing entirely new business. Beyond the low-cost sector we are seeing a new breed of Indian carrier, such as Kingfisher, capture a totally new market with a focus on quality products. The emergence of Middle East carriers such as Etihad and Qatar Airways are good examples of opportunist behaviour in freshly emerging markets.

In the case of a start-up challenging an incumbent, the stakes are very much higher because a start-up is faced with immediate competition from an operator with, presumably, an established reputation and, more importantly, deeper pockets. An example of where things can go horribly wrong is the first incarnation of Compass Airlines, which launched itself into a newly deregulated Australian domestic market in 1991, only to fail in 1992. Compass' decision to serve the main trunk routes of Melbourne-Sydney-Brisbane brought it into immediate and head-to-head competition with both Australian Airlines and Ansett Australia.

Here we need to introduce another element in the growing complexity of airline market dynamics: the question of frequencies *versus* capacity. Compass suffered for numerous reasons, including under-capitalisation, passenger terminal access limitations, and problems with telephone call centres. Crucially, they were unable to build the frequencies they needed to establish themselves in business-dominated markets. Their A300-600R aircraft were inherently economic and efficient, but were oversized for *their* particular market strategy. In a market largely dominated by smaller A320 and 737 units, it was always going to be tough-going for any newcomer to carve a niche.

An Alliance Objective

Closely linked to the development objective is that of developing a strategy of alliances. The character of the strategic alliances has changed somewhat since the initial days of the 1990s. It is becoming more evident that cost synergies are actually quite difficult to realise, although revenue enhancement is quickly measured. However, many of the advantages of a merger can be achieved through alliance partnerships. The major strategic alliances are certainly here to stay, and even though most of the world's major airlines have now made their decisions, 'alliance churn', or rollover of membership, is likely to remain a possibility. With alliances as popular as they are, it is important to reflect on how they should be considered in the attributes of a fleet plan.

There are many different types of alliance, ranging from the simplest forms of technical co-operation and code sharing, through to virtual mergers of operations, with the exchange of equity and sharing of costs and revenues. The success of such groupings is improved where airlines have common fleets and standards. For example, the maintenance of Spanair's A320 fleet by Lufthansa is a very logical decision, given Lufthansa's considerable expertise with that aircraft family.

As we saw in Chapter 1, today's major alliances are driven by two principle objectives: improve market reach, and therefore revenues; and to reduce costs. It is probably true to say that the first of these objectives is easier to achieve and will probably preoccupy airline partners in the first few years of a relationship. Beyond, opportunities to save cost must certainly be present; although achieving meaningful savings in large investment areas will prove insurmountable unless partners are willing, and able, to compromise.

An Economic or Financial Objective

The fourth airline objective is less connected to the type of aircraft or type of operation *per se*. The aircraft fleet is a means to an end, rather than an end in itself. It is a vehicle for a business to achieve its objectives. Most airlines today strive for profitability rather than the fulfilment of a public service obligation. Given that this is the case, a set of economic or financial criteria needs to be determined.

A profitability goal may be short-term or long-term. Start-up airlines are rarely profitable in their first year of operation owing to the degree of initial costs in order to become established. Under-capitalisation is a common reason for start-ups to fail. Even a $50 million budget may be hardly sufficient to get over the hurdle of the first year. The choice of aircraft can determine the financial success of a start-up. As we saw with Compass Airlines, a wrongly-sized aircraft contributed to its downfall. On the other hand, New York start-up JetBlue's choice of A320 turned out to be a prudent choice as the economics and market appeal of the aircraft were recognised. AirTran committed to the 717 but then found itself operating a large number of a type that ultimately became obsolete. Unlike start-ups, established airlines do not have the concerns of achieving early profitability in order to merely survive. Yet, they must also satisfy the financial criteria set out by their owners or shareholders.

So where does the composition of a fleet affect the ability of an airline to meet economic and financial objectives? Managing the number of aircraft owned or leased can control the magnitude of investment in a fleet. Off-balance sheet leasing structures can play an important role in managing the debt-equity ratio of a business. Also, the investment appraisal of an aircraft acquisition project should reveal whether return-on-investment targets are going to be achieved, and over what period. Risk assessment in an appraisal also plays a vital role.

In fact, there are many strategic reasons why managing the proportion of owned and leased aircraft is important. For example, a basket of aircraft from two or more suppliers can enable a degree of leverage. This may extend to the suppliers of the engines and other equipment as well. Conversely, there may be financial advantages to pursue a single-supplier philosophy, although whether this is ultimately in the best interests of an airline will depend upon the diversity of the supplier's product range, support package and overall long-term market position. Airlines that made heavy investments in Fokker or McDonnell-Douglas products face an increasingly challenging set of circumstances today.

So, What Does Make a Good Fleet Plan?

There are three basic attributes, which can even be considered golden rules, of fleet planning: adaptability, flexibility, and continuity.

Attribute 1 – Adaptability

Success in an airline can be measured in many ways, profitability being the most obvious. However, to achieve a result the business has to manage both its supply and its demand. This is not as easy as it sounds. Neither supply nor demand are totally within the bounds of control of airline management. Supply decisions ought to be driven by the need to meet objectives and goals, yet circumstances often conspire against the supply conditions that an airline would ideally prefer. Limits of capital, staff, expertise, and aircraft availability may dimension the amount of supply generated. Regulatory constraints, ranging from environmental restrictions to slot limitations, may also control the amount of supply an airline can place into a market.

The other side of the equation is equally difficult to control. Target demand is one part of the total air travel demand it has elected or is authorised to satisfy. An airline may not be able to capture all of the target demand, especially if it is in competition with other airlines and, sometimes, other modes of travel. The share of the demand that can be captured is largely a function of the airline's own objectives, pressure exerted by the competition and the degree to which supply is adapted to the demand. The airline brand, pricing and distribution strategies, and the mere presence of capacity in the market can heavily influence demand at the right time.

Assuming that an airline can positively influence both its supply and demand, then the path to success is in bringing the two elements as closely together as possible. Yet, even an ideal matching of supply and demand is no guarantee of profitability.

There are some specific areas that can be identified in the context of fleet planning that enable us to measure adaptability.

Does the aircraft have the right size and appeal?   Aircraft capacity is firstly a function of the magnitude of the market addressed. If the strategy is to serve high-density trunk routes, then high-capacity aircraft, with consequent economies in terms of unit costs, can be considered. However, as we have seen, frequency-sensitive business markets often favour frequency over aircraft size, so smaller aircraft might be more appropriate. If the strategy is to by-pass a hub, either because delays at the hub may be excessive, or because a niche market is pursued, then this might also suggest smaller aircraft size as routes might be expected to be of lower density.

There is a competitive issue to be considered. Airlines that serve very small home markets might be forced into smaller capacity aircraft, whereas they could find themselves competing against big carriers having the critical mass to employ large fleets of large aircraft.

Linked with the size of the aircraft is comfort. Clearly, large aircraft offer more opportunities to develop and exploit the brand. Decisions have to be taken concerning the space allotted to each passenger. Seat pitch, being the distance between seat rows, and the number of seats abreast is a contentious issue in aircraft economic comparisons. Although comfort may be regarded as a subjective element, it has significant repercussions when dealing with economics. Packing more seats into the tube does wonders for unit cost, but can significantly diminish the attractiveness of the product in a competitive market. The low-cost carriers do not worry too much about this. Indeed, the emergence of 'slim' seats provides a good compromise between comfort and seat density. At the other end of the scale, Singapore Airlines' A340-500 operation from Singapore to Los Angeles and New York is based on a two-class configuration of only 181 seats, compared to a manufacturer's three-class standard of 313 seats. Although this low seating density provides spectacular comfort levels, it also has a dramatic effect on the unit operating cost. It must be accepted that there are performance considerations that would limit seating density on these very long routes. Also, the airline is able to extract a premium price for the service which helps to offset the apparent penalty.

Aircraft size and appeal should be considered not only for today's needs but for tomorrow's as well. Markets that are volatile in nature, or that are expanding at a rapid rate, are difficult to service because the aircraft never seems to be the right size at the right time. To a degree, it is possible to close-up seat pitch and provide more effective capacity within the same airframe, but this solution runs against the trend for airlines that are thinning-out the number of seats.

A key issue in appeal is that the airframe should be able to offer a good degree of flexibility for reconfiguration. This is more easily achieved with large twin-aisle aircraft than with small types. Different cultures and evolving passenger needs dictate seating arrangements and the overall feel of a cabin. So, airlines operating in markets that generate the density of demand that require large aircraft have the ability to be responsive to market changes.

The ability to reconfigure an aircraft quickly is an absolutely key attribute for a leasing company. Lessors may need to place the same aircraft three or more times during its life. Minimising the complications of reconfiguration therefore becomes very important.

Does the aircraft deliver the right performance?   Traditionally, aircraft selection decisions were dominated by the range capability of the products under evaluation. This was because range was often a limiting factor for airlines wanting to expand their networks. Today's large civil aircraft have greater range capabilities and we have seen this element diminish in significance. Indeed, current offerings from both Airbus and Boeing include aircraft with a capability of well over 8,000 nautical miles (nm) range with a maximum passenger load. This means that the vast majority of the world's city-pairs that can generate sufficient demand can be served non-stop. It is unlikely that aircraft will be developed with significantly more range ability, as performance limitations tend to become exponential in nature. Greater distance requires more fuel, and more fuel is burned in order to carry the extra fuel to achieve the range.

The 'holy grail' in terms of range is to fly non-stop from Sydney to London and back. Both the manufacturers have tried to satisfy this requirement for Qantas, but the prize of an order for an aircraft to perform this mission just keeps slipping away.

Another significant development in terms of range concerns the geographical positioning of the major hub airports in the Gulf region. Today's longest range aircraft, such as the 777-200LR and the A340-500 are capable of flying literally to any airport on the planet non-stop from Dubai. This explains why airlines domiciled in this region anticipate turning airports in this zone into the crossroads of the world, providing connecting services to practically anywhere. However, they do need aircraft to enable this bold perspective.

Many airlines do place enormous importance on range when comparing aircraft types. There are some sound reasons for this. For example, the future value of an aircraft is partly determined by its range ability. Operators of very early models in a production run can often be disadvantaged, as take-off weights, and therefore range, are always limited until improvements can be worked into the design. It is very common indeed for the first version of a type to be offered with a somewhat conservative payload-range envelope. This is largely because the structural integrity of the airframe must be proven. In fact, loads on an aircraft structure are governed not only by weight, but also fatigue life and static load. Limiting factors also vary according to the number of engines and centre-of-gravity position. As experience with the airframe grows, so does the design engineer's confidence in releasing increases in take-off weight that had been retained as margins in the original structural design.

These take-off weight increases can be of some magnitude during the life of an aircraft programme. Even in the design and development phase, the MDTOW may start off at a value too low to maximise the sales potential of the aircraft. Hopefully, the first delivered aircraft have evolved to a reasonable MDTOW. During the life of the aircraft design in service, one might expect several evolutions of the weight, each one enhancing the maximum range possible. Such developments are often accompanied by many other changes to the design as well, such as increased fuel volume and other product enhancements.

As an example, the A340-300 take-off weights evolved from 232 tonnes up to no less than 275 tonnes during the development and evolution of the aircraft.

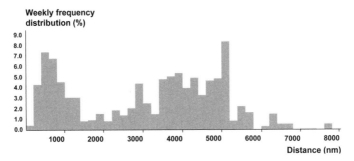

Figure 2.1   Flight Distribution by Distance – A340

What is important is not so much the inherent range ability of an aircraft, but whether the type can deliver the range required for the network under consideration, both today and in the future. Clearly, it is difficult to predict with certainty the kind of route structure likely to be flown at some stage within the life of the aircraft in the fleet, so some conservatism is always worthwhile. Acquiring an aircraft with too much range means higher costs in areas such as landing fees and maintenance costs.

Ideally, there should be a match between the stage lengths in the airline network and the optimum range of the aircraft employed. In practice, this hardly ever happens. Route networks tend to be heterogeneous, comprising a mixture of stage lengths. Even airlines that can be described as either 'long-haul' or 'short-haul,' can find themselves with a variety of route lengths. The problem is that each and every aircraft type has an optimum design range that describes the maximum range over which it can carry a defined amount of payload (usually the maximum passengers or the maximum passengers plus cargo). As we shall see in Chapter 5, the payload-carrying performance of the aircraft will degrade beyond this particular range. Similarly, at ranges less than that achieved with the maximum number of passengers or payload, the productivity of the aircraft is not being maximised. The fleet planning problem, therefore, is one of reconciling a basket of routes, comprising perhaps a wide variety of stage lengths, with an aircraft's basic payload and range design.

It is inevitable that aircraft are operated away from their optimum point. Cathay Pacific once referred to this phenomenon as 'intelligent misuse,' which brilliantly sums up the problem.

Figures 2.1 and 2.2 show the number of weekly frequencies scheduled by operators of the world fleet of A340 and 777 aircraft in 2004. The achievable ranges by both aircraft well exceed 8,000 nm, according to aircraft configuration and operating rules. Although these are clearly very long-range aircraft it is often operationally sensible to use the aircraft in addition on 'tag-end' sectors to long-haul flights, or else much shorter routes in order to boost utilisation and revenue opportunity. Flexibility to deploy the aircraft on a variety of routes is a key attribute.

It follows that the performance and economics of any aircraft can only be correctly assessed against the *actual* network operated and there may be some differences in aircraft capability when it is employed on sectors that are far from those designed for the aircraft. This is what intelligent misuse is all about.

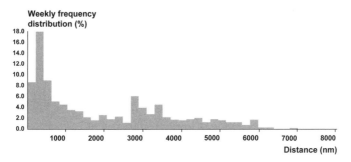

Figure 2.2   Flight Distribution by Distance – 777

One type of airline that is less concerned with flying aircraft away from their optimum range is the low-cost carrier. The philosophy of a traditional airline is to organise its schedule and aircraft deployment so the market is best served. In other words, the airline tries to serve its market. However, the low-cost carriers know very well that their market is much more price-sensitive and less inclined to object to inconvenient scheduling. So this breed of carrier has the opportunity of optimising its aircraft utilisation and, in a sense, forcing the market to move towards the schedule, rather than the other way around. This is one reason, although clearly not the only one, why low-cost carrier economics are superior than those of traditional airlines, even though they may operate the same aircraft type.

Does the aircraft deliver the right economics?   As we shall explore in Chapter 6, economics is not concerned with costs alone. It is concerned with costs and revenues. In deciding whether or not the fleet plan is adaptable, an evaluation of the operating costs is clearly essential. Great debate always surrounds the composition of an aircraft cost breakdown. Achieving the 'right' economics is rather like achieving the 'right' performance. In other words, 'right' is not necessarily 'least'. As we have just seen, using an aircraft with the maximum range may not be the most appropriate solution if that range is not strictly required for today's or tomorrow's network. The same can be said for economics. Of course, it is in the interests of the airline to minimise costs, but there are some additional considerations that might result in higher costs being incurred in order to ensure a better overall result in the longer term.

When the rules governing the operation of twin-engine aircraft further from 60-minutes' flying time from diversion airfields were first promulgated, operators were rightly concerned about the investment required in order to adhere to these ETOPS (Extended Twin Operations) rules. In the mid-1980s, they wisely predicted additional investment in maintenance costs in order to ensure that the reliability targets set down in the ETOPS rules could be achieved. Other areas of ETOPS costs abound, such as training costs, spares investment in order to protect a more stringent Minimum Equipment List (MEL), and the costs associated with a diversion that may not be necessary for a multi-engine aircraft. ETOPS is highly dependent upon the region of the world being served. Now that ETOPS has become so commonplace the early investment made by operators has now paid off in terms of better reliability, enhanced operational practice throughout the operation and, ultimately, lower costs.

When making the crucial decision concerning the number of preferred engines on the airframe, the airline should balance the operating cost against the differing investment in spare engines.

The right economics may also be judged according to how forgiving the aircraft type may be in certain areas of the cost breakdown. For example, if an operator is serving airports where landing fees are relatively high, then it might pay to throw more emphasis on the weight of the aircraft in the evaluation. Some aircraft types are more efficient, or productive, in terms of weight per unit of payload than others.

In an environment of perpetually high fuel costs, the emphasis has clearly returned to the fuel efficiency of the aircraft. This had previously occurred in the early 1980s, which coincided with an earlier phase of high fuel price.

Fleet planners are continually faced with the problem of how to predict whether today's cost drivers will be tomorrow's.

Attribute 2 – Flexibility

Flexibility has become the watchword in airline fleet planning. Here we are faced once more with resolving the dilemma of matching a physically inflexible aircraft with markets experiencing constant change.

Does the aircraft design offer sufficient payload versatility?   We have already discussed that long-haul aircraft, because they tend to be large and with twin-aisles, tend to offer a greater degree of flexibility than small, single-aisle aircraft that serve local or regional markets. Greater interior floor space and volume tends to mean more opportunities in terms of seat layout, positioning of toilets, galleys and other amenities, design of overhead stowages, provision of facilities for cabin and flight crew and so on. A key question is whether the inherent design of the cabin allows cost-effective and rapid reconfiguration in order to address a market change.

The issue of payload versatility can be seen from the perspective of the longer-term design for the market, or else as a day-to-day operational issue. Airlines are far more responsive to market changes than ever before, with regular reconfiguration of the cabin environment now commonplace. A complete image revamp for a fleet can easily involve costs well beyond that of a new aircraft. So, a good fleet planning objective is to ensure that the cabin of any aircraft type can be altered rapidly and economically. Whether or not good flexibility is achieved is partly dependent upon the initial specification of the aircraft. For example, it may be a good investment to specify the maximum number of seat rails so that, at some stage in the future, a high-density charter configuration can be installed.

Another aspect of flexibility that should be considered, is the degree to which the aircraft design offers alternative loading options. For example, it is always beneficial to be able to load full-sized pallets in a transversal sense in both the forward and rear cargo holds. An inability to place pallets in a rear hold can restrict market opportunities.

Does the aircraft family offer sufficient operational flexibility?   The most important ingredient in determining the degree of flexibility, and one to which we shall return many times, concerns that of multi-sized and multi-range versions of the same aircraft family. Boeing, Airbus and the former McDonnell-Douglas all developed families of aircraft. Building up a family of aircraft of varying sizes, but sharing overall design and operational commonality, should be at the heart of any successful fleet plan requiring different size modules to serve different markets.

The family concept has had a dramatic effect on fleet planning as airlines have reaped enormous economies in terms of both investment as well as operation. Where a sequence of aircraft types is offered with designed-in commonality, a fleet can be better matched to expanding demand because capacity can be added by trading-up in size as well as in units.

Operational flexibility becomes especially important in the fleet plan as future range requirements can be more easily planned in advance. There is much less risk involved in making an aircraft selection where it is known that solutions to future growth can be found by simply moving-up to a larger-sized variant of something you have already got.

Also, the ability of the manufacturer to propose multi-range versions of an aircraft type can help the airline better match the aircraft type to particular parts of its network.

Does the fleet plan allow easy phasing-in or out of capacity?   A final flexibility attribute concerns the degree to which the timetable of aircraft deliveries can be modified to reflect changing conditions. Airlines are enormously affected by external factors, ranging from competitors' actions and changing macro-economic conditions, to new regulations and even war. The lead-times between the placing of an order and taking delivery of an aircraft are necessarily long, due to the sheer complexity of the product and its support in an operation. As we all know, circumstances can conspire against the best-laid plans of mice and men.

Type selection within a family of aircraft is an important form of flexibility. Thus, the newly-merged US Airways and America West were able to alter their fleet mix on order to address evolving needs. As well as placing new orders, they migrated eight existing orders for A320s into A321s, as well as extending conversion and cancellation rights with respect to 15 A318s. The company also advanced the delivery schedule by one year to 2008. This transaction provided a huge degree of flexibility for US Airways.

If the manufacturer can offer a degree of flexibility between the number of firm orders and options, then this can go some way to helping the airline adapt to changing needs and uncertainty. Converting option commitments into firm sales is obviously a manufacturer's sales objective.

Attribute 3 – Continuity

As time marches on, we want to be sure that incremental changes to the amount of capacity in a fleet can be made with minimal disruption to the operation.

Is the product line coherent?   Product coherence can be considered in two ways. Firstly, we need to judge the degree to which each member of a product line acts as a point of reference for the airline. For a supplier to break the monopoly of an aircraft or engine type in an airline is extremely challenging. We can easily expect different philosophies in design and maintenance, and different approaches to customer service and management in general. There have to be good reasons to break the mould. Of course, airlines may pursue a dual-supplier philosophy for strategic reasons, such as spreading risk and encouraging more competitive offers from manufacturers. Perhaps a dual-supplier policy is necessary because neither manufacturer offers competing products, with Boeing's dominance of the 747 market being a case in point. In general, only large airlines can economically muster the internal resources to support operating multiple types.

A second aspect of product coherence concerns the composition of the product line itself. Airframe and engine suppliers have striven to build broad product ranges so that plenty of choice is on offer. Both the Airbus and Boeing product range address markets needing short-haul 100-seater aircraft through to markets requiring high-capacity long-haul types. Yet what really counts is not the presence of a wide variety of different aircraft types in the manufacturer's portfolio, but the economic fit of different members of a family.

Clearly, once a decision has been taken to acquire a particular aircraft type, then it is logical to stick with that particular technology standard as the fleet develops. Besides the obvious benefits of economies of scale and quick progression down the learning curve, there are synergies to be reaped in all aspects of the customisation process, spares investment, tooling, training, and day-to-day operations.

Key Decision Criteria

It goes without saying that each decision process is unique. Although there is no magic recipe that can be applied to fleet planning, a structured and prioritised set of key criteria, relevant to the airline and its position in the market, is essential.

Timing is Everything

Importantly, these decision criteria may evolve over time. For example, in the early 1980s the price of fuel was relatively high whereas interest rates were low. Against such an economic background fuel efficiency would clearly assume greater prominence than aircraft first price. Yet by the end of the decade fuel prices had fallen and interest prices had risen, thereby influencing the ranking of decision factors. The roller coaster continued when, in the mid-1990s, fuel price remained low as the cost of borrowing reduced. By middle of the 2000s, fuel price skyrocketed, throwing fuel efficiency once more into sharp focus.

Sometimes it is the availability of the aircraft at a moment in time that determines choice. The series of iterative loops that Airbus passed through in the definition of the A350 clearly handed an advantage to Boeing, who could bring their competing product, the 787, to market several years earlier. The difficult question to resolve is whether an airline is able to wait for the better product and what kind of impact this may have over the longer term. In the case of fleet renewal, the airline faced with this choice may need to delay the retirement of existing aircraft. This could be quite punishing for airlines in desperate need of fresh technology. However, for airlines seeking to expand into a new market, the decision is slightly less complex.

The Effect of Technology Exhaustion

In the days of economic regulation of the airline industry, the pure cost of operation mattered a great deal. Consequently, decisions for aircraft were mostly driven by the unit-cost advantages. This was particularly true of the 747, for example. Deregulation and the growth of competition saw the emphasis shift to revenues and the ability of the aircraft to contribute to the brand of the airline. Now we see a definite trend whereby passengers see the airline product more and more as a commodity, despite the best efforts of quality-driven airlines to convince us otherwise. Airline choice decisions are driven by elements such as the schedule and on-time performance, with brand loyalty sometimes taking a back seat. Such a trend deflects the technology of an aircraft from a position of prominence in the list of decision criteria.

Another reason for technology assuming lesser significance is that manufacturers are approaching the limits of what can actually be achieved through conventional technology. Incremental improvements in fuel efficiency are becoming increasingly difficult to realise. This is rather akin to the progression of sporting records. When measurement and competition was in its infancy the records tumbled quickly. It is now increasingly rare to see records broken as the limits of human endurance are approached. Attention is now turned to nutrition and training in order to make further improvements to sporting records.

So if technological advance becomes exhausted, where do we turn? The answer may lie in better optimising the deployment of aircraft and better matching of demand to capacity.

Range evolution of aircraft has probably reached a plateau as the majority of city-pairs that can generate sufficient traffic to warrant the development costs to achieve vast ranges are now linked up.

Technology has become much more reliable than ever. Forty years ago engines spent around 1,000 hours on the wing. Today we would not raise an eyebrow if an engine achieved 20 times that. How much better can reliability become at reasonable development cost?

Yesterday's technological breakthroughs are now considered the norm. All-weather operation, inter-continental ranges and two-person flight decks have all ceased to be marketing tools. However, there is still scope for some technological advance. We will continue to see the application of new materials and aerodynamic improvements, but these will have a lesser impact than previous technologies.

It has to be accepted that technology will one day no longer drive airline operating costs.

The Impact of External Forces on Decision Criteria

It is clear that in a world of sweeping liberalisation, markets are no longer stable, and we have already examined the problems of matching inflexible supply and volatile

Relative value of criteria

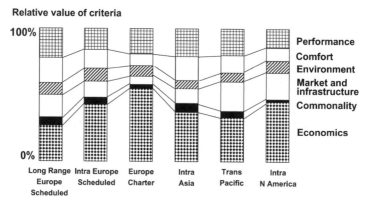

Figure 2.3   Key Buying Factors
Source: DASA

demand. Other regulatory changes can also significantly affect fleet planning. The airline industry is an obvious concern to the environmental lobby. Airlines finding themselves in the thick of the environmental debate have no choice but to elevate the issue to a key decision criterion. Airlines serve airports that are increasingly impeded by environmental restrictions and the subject is becoming more and more a dominant theme in aircraft design and fleet planning. For example, the 'QC2' noise restrictions at London's Heathrow Airport actually led to a review of the design of the A380.

Airlines may fall victim to exchange rate evolutions that may completely overshadow any other cost parameter. The weakening value of the local currency, in which revenues – but not costs – will be generated, might well be the most potent issue in fleet planning for some.

Geography also plays a role. Forces at work in one part of the globe will not be apparent in others. For example, there is a tendency for Asian markets to expect twin-aisle aircraft, even for short missions. The environmental lobby is a much less significant issue in developing economies, unless airlines domiciled in such regions aspire to serve markets in countries where environmental considerations apply.

Fixing the Key Decision Criteria

We have just concluded that key decision criteria are unique to a particular situation, are highly sensitive to timing, are becoming less and less contingent upon technology, and are governed by external factors such as regulations, the environment and geography. Fixing a set of decision criteria is obviously not easy and there are many ways of skinning this particular cat.

DaimlerChrysler Aerospace (DASA) studied key buying factors of airlines world-wide to help establish aircraft design objectives and provide economic and operational targets for airlines and manufacturers (see Figure 2.3). Their studies emphasised the importance of aircraft design features in addition to more classical economic advantages. Not only did they establish that more than half of these key buying factors concerned elements other than economics and commonality, they concluded

Category	Elements
Markets and routes	Size, growth, mix, comfort, schedule, Airport compatibility, economics, turn times
Operations	Crewing, aircraft mix, ETOPS, Minimum Equipment List, performance
Finance and contractual	Purchase vs lease, residual value, buy back, insurance, price escalation, guarantees, spares pricing, cost of updates
Engineering	Spares inventory, pooling, commonality, facilities, third party needs
Regulatory and environmental	Certification rules, environment standards special conditions

Figure 2.4   The Fleet Selection Process

that so-called added-value factors were increasing in significance. What singles out their study is how they related the results to different geographical markets.

Perhaps unsurprisingly, economics is the single most important element in their analysis irrespective of geographical region. Aircraft performance scored very highly for European-based scheduled long-range operations. Comfort is also a preoccupation for this type of operator, unlike in the United States. The lower score assigned to commonality in the US is due to the comparatively larger fleets where economies of scale are more easily present. In other regions, where fleet sizes are much smaller, there are more potential synergies available through common aircraft technologies.

By blending aircraft-specific buying factors with those connected with the operating environment, DASA's study has shown the decision-making process in a fresh light. However, recommended practice is for airlines to compile their *own* hierarchy of key buying factors relevant for both today and for the future. Individual airlines within the regions studied by DASA may find completely different results.

However, the DASA study is but one method that could be applied. Another approach would be to throw everything into the pot and build an overall picture of all of the elements that come into play. In Figure 2.4 we can observe an alternative way of seeing the challenge.

This approach involves defining a series of categories that are directly affected by aircraft type. Against each one is a non-exhaustive list of elements that would need to be examined and, importantly, prioritised.

There are many ways of separating elements of an operation that are directly linked to the aircraft. Another methodology involves listing a set of criteria in order of current and future importance, as seen in Figures 2.5 to 2.7. Each Figure shows a sample set of criteria, with the elements in the 'Must have' columns arranged in sample orders of priority. Note that in each of the three cases the same aircraft

Low-cost carrier	Must have	Needed	Nice to have
Low operating cost	1 ✓		
Rapid turn-round time	2 ✓		
High reliability	3 ✓		
Good field-length performance			✓
Early aircraft availability		✓	
Compatibility with other aircraft in fleet		✓	
Large overhead stowage volumes			✓
Easy, rapid loading procedures	4 ✓		
Simplified cabin definition	5 ✓		

Figure 2.5   Aircraft Selection Criteria – Low-Cost Carrier

might be under evaluation, but the elements that may be considered important are completely different.

Compiling such a set of criteria is really an essential exercise before the aircraft evaluation work is started. Referring back to the chart once the final decision has been made may be a salutary experience. This approach is helpfully performed with the manufacturer. In this way it is possible to find common ground as to where to focus attention and resources.

As we shall see later in the book, much of what we have discussed so far in this Chapter can be calculated. There is a school of thought that says that fleet planning should be one-third calculation and two-thirds based on confidence, judgement and trust. We should not fall into the trap of calculating things which simply cannot be calculated and should avoid over-analysing scenarios. The complexities of markets are

Short-haul scheduled carrier	Must have	Needed	Nice to have
Delivers optimum economic value	1 ✓		
Future design development potential			✓
Competitive passenger appeal	2 ✓		
Easy to maintain		✓	
Spares pooling ability		✓	
Ability to differentiate in the cabin	3 ✓		
Excellent customer perception	4 ✓		
Containerised freight and baggage	5 ✓		
Good working environment for crew		✓	
State-of-the-art flight deck and avionics			✓
Extensive technical assistance available		✓	

Figure 2.6   Aircraft Selection Criteria – Short-Haul Scheduled Carrier

Lessor	Must have	Needed	Nice to have
Generous pricing from supplier	1 ✓		
Highly regarded by the market		✓	
Adaptable to a wide range of markets		✓	
Available with a choice of powerplant		✓	
Retains value	2 ✓		
Possibility to extend life as a freighter			✓
Low reconfiguration costs	3 ✓		
High degree of component standardisation	4 ✓		
Good operational flexibility		✓	
Unlikely to be technically obsolete soon			✓

Figure 2.7   Aircraft Selection Criteria – Lessor

such that any decision is going to be both a compromise and something of a guess. Of course, decision-makers need the comfort of a well-researched and analytical study.

Nevertheless, we would do well to recall the principle of economy of thought promoted by the fourteenth-century Franciscan philosopher William Ockham. His ancient maxim, called Ockham's Razor, says that what can be done with fewer is done in vain with more.

When it comes down to it, fleet planning is just as much a human exercise as a computer-driven one.

Trends in Tools and Data

Early fleet planners were plagued by both a lack of sophisticated tools and data of any usable kind. Tools were initially limited to analysing the performance of the aircraft. Emphasis was traditionally placed on network 'hot-spots,' where an aircraft would be limited in terms of its take-off ability from a particularly hot or high airfield, or else restricted in payload on excessively long missions. Data would be limited to passenger boarding by sector, plus rudimentary cost models. Before airlines began to compete vigorously, it was even unusual for revenue generation of differing aircraft types to enter into the fleet decision at all.

Today's fleet planner has at his disposal a wealth of decision support tools and data – probably too much. Let's consider how they have evolved.

Decision support tools for fleet planning have always been dedicated to that purpose alone. Here we must make a clear distinction between tools intended to evaluate the best fleet mix over a period of time, and operational tools which determine the best *use* of an existing fleet. The purpose of fleet planning is to find the best combination of aircraft and aircraft types to serve a certain market demand. Such tools tend to be long-term in outlook and look for an optimum solution. There is no particular virtue in integrating, for example, a scheduling module or detailed maintenance plan into such a long-term outlook. Such elements are more suited to an operational approach.

Data sources have also changed considerably over the years. There was a time when the analysing airline, or 'host' airline, only had its own market and traffic data at its disposal. Today, it is possible to gain access to large amounts of total market data, through Marketing Information Data Tapes (MIDT). These data comprise bookings amassed through a certain number of Computer Reservation Systems (CRS). Access to these data comes at a price, and only through a small number of processing agents, who are responsible for the 'cleansing' of data, such as the removal of personal information, for example. Also, these agencies can customise the presentation of data. The ability of a host airline to have access to booking data of its competitors has had a tremendous influence on its ability to compete. Total market knowledge can give an edge in planning new routes, measurement of the success of product changes, brand strategies, and the appropriate mix between frequencies and aircraft size.

Knowledge of booking levels by flight number and booking class can prove invaluable in tracking market evolution and competitors' actions. However, it should be borne in mind that the data available through MIDT is limited only to traffic that has actually booked through a Computer Reservations System (and one which releases its bookings through this medium as well). Another disadvantage is that the data comprises bookings, rather than final on-board loading. Thus, the data do not suggest any shifts between recorded bookings before a flight departs. Data are depleted in certain markets of the world, Asia in particular. Also, MIDT data cannot reveal bookings taken directly by an airline or bookings made through the Internet, rather than through a CRS. Despite these shortcomings, MIDT data are certainly very useful in tracking trends in a market, even though absolute values may be inaccurate.

The availability of such data in the public domain has been a boon to everyone in the industry seeking, and willing to pay for, broad market knowledge. However, as more and more bookings are made on the Internet it has to be accepted that the value of MIDT as a source of data for fleet planning is now diminishing significantly.

Another type of data that is vital in fleet planning is revenue management. The valuation of revenue gained from each seat on an aircraft plays just as important role in fleet planning as the cost of providing each seat. Incredibly, it has taken a long time for the fleet planning community to embrace revenues as part of the economic equation. It is very difficult to locate accurate detailed information on airline yields because these data are regarded as highly confidential. However, overall values and trends are often published in airline reports and trade associations.

In truth, the most accurate and useful data for fleet planning purposes will be internally generated by the airline. The trick is in managing the huge volumes of data to avoid becoming swamped.

In Summary

In this Chapter we outlined a set of value attributes that may apply to the study airline. In this way we can judge the degree of importance attached to elements such as the network, the staff, the brand and the fleet itself. We then examined various overall goals and objectives in order to see the extent to which the fleet may make a contribution to their success.

All of this knowledge goes on to assist in gauging the attributes of a fleet plan. Elements such as adaptability, flexibility and continuity are watchwords for any fleet plan.

Airline structures and objectives are extremely diverse, but it is essential for fleet planners to rank both overall buying factors and aircraft-specific selection criteria.

We shall now turn our attention to the aircraft evaluation process itself, which divides fairly neatly into five broad steps. Firstly, we need to develop a good understanding of the market in which we intend operating. Secondly, we need to look closely at the aircraft product, to ensure that it can enable overall airline objectives to be achieved. Thirdly, we must analyse aircraft performance to ensure that the options under consideration can carry the requisite payloads on the routes. Then, we should evaluate the economics of the operation, and finally conduct an investment appraisal to examine the operation over a period of time.

Let's start with the market evaluation.

Chapter 3

The Market Evaluation

Setting the Scene

In this Chapter we shall address the fundamentals of market modelling as applied to aircraft decision-making. The first step will cover the basics of productivity measurement. In order to facilitate a comparison between different types of market model, we shall construct a simple route network. We will then examine two seemingly opposing, yet complementary, approaches to the modelling of demand: the macro and micro approaches. With reference to our route network we shall firstly construct a macro fleet plan. Then we shall address market segmentation and its link to the all-important concept of spill, which is defined as an excess of average demand over capacity. We shall look at a modelling approach appropriate for a single sector, then expand this into a more complex micro-network model. Finally in this Chapter, we shall take a look at some specific and independent techniques for assessing market share.

The concepts we shall examine relate very much to the scheduled, rather than charter business. The economic cycle, as well as hard-to-predict shifts in demand drives the scheduled market over time. On the other hand, the charter market, whilst not immune to economic cycles and seasonal variations, is nevertheless resilient, stable, growing and predictable.

In order to inject a sense of direction and consistency into the discussion, we shall consider the exploits of three mythical airlines competing in the same market.

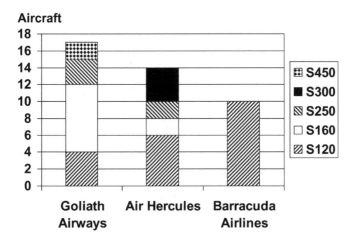

Figure 3.1   Fleet Compositions – 2007 (Aircraft)

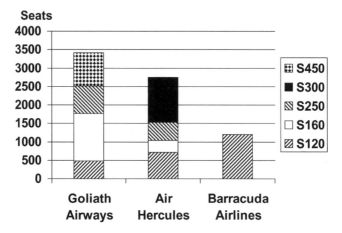

Figure 3.2   Fleet Compositions – 2007 (Seats)

Two incumbent airlines have been operating for a number of years in a stable and regulated environment. They are called Goliath Airways and Air Hercules. Deregulation is on the verge of implementation and newcomer Barracuda Airlines will enter the fray. Goliath and Hercules are bound by their existing fleets, which are composed of multiple types and are not necessarily optimal, but Barracuda will enter the market with a single aircraft type.

The fleets of the three players are shown in Figures 3.1 and 3.2. It is always a good idea to visualise aircraft fleets in terms of the number of units and the number of seats, as the relationships will differ. The two incumbent airlines are saddled with four aircraft types, of widely varying sizes. This may be advantageous if demand varies a great deal in magnitude over time. However, there may be serious economic challenges due to the small fleet sizes, unless significant commonality exists between the types.

The network on which Goliath Airways and Air Hercules currently compete comprises one significant trunk route, from the hub of Metropolis (code MET) to Neighbourhood (code NEI), and a basket of smaller routes serving regional destinations. Barracuda Airlines will compete on the business-driven MET-NEI route. The network is seen in Figure 3.3. Owing to the significance of the Neighbourhood market, this single route merits being considered as a sub-network all on its own. The second sub-network comprises all of the other regional routes. The purpose of dividing the operations is two-fold. Firstly, where one route is clearly dominant in a route structure it makes sense to isolate it in order to better judge the degree to which its performance affects the entire operation. Secondly, a major route may possess specific characteristics that could differ from those of the rest of the network.

In the example, as in real operations, aircraft are not constrained to any particular route or sub-network. Also, in the interests of clarity we shall conduct the analysis on the basis of a typical week. However, it would be usual to consider cyclical variations in both demand and supply. One of the major problems that we will have to face is that where demand does vary considerably by time period it is not practical to keep changing aircraft assignments in order to achieve near-perfect capacity matching. Compromise is almost always necessary.

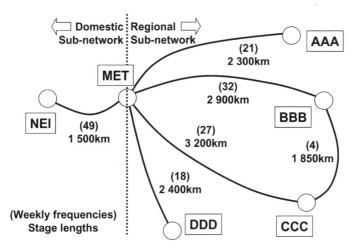

Figure 3.3   Study Network

In starting our fleet plan we will need to use some basic measures of production, so we will briefly turn our attention to defining the main measures that are central to airline operations and planning.

Measuring Production

Defining Production and Sales

A logical start to the fleet evaluation process is to make an assessment of the degree of productivity of the fleet, either for the airline under study, or for the whole market. We shall do this by aggregating Goliath Airways' production for the Metropolis to Neighbourhood route, measuring by Available Seat-Kilometres (ASKs) and Available Tonne-Kilometres (ATKs). One ASK equates to one seat being flown 1km and one ATK represents one tonne being flown 1km. The corresponding Revenue Passenger-Kilometres (RPKs) or Revenue Tonne-Kilometres (RTKs) are indicators of how much of the production has been sold.

It is quite common to convert all seat production and sales into ATKs and RTKs, by assuming an average weight per passenger and baggage. Once this conversion has been performed the amount of freight carried in the underfloor can be added to these values so that the entire production and revenue capability of the aircraft are considered.

Another very important measure is the Revenue per ASK, or RASK. This is a combination of the yield and the load factor. Thus, the ratio between ASK and RASK gives an indication of both cost control as well as revenue generation.

Defining Load Factor

The ratio between ASKs and RPKs, or ATKs and RTKs, is the load factor. We can establish from the data in Figures 3.4 and 3.5 that the Goliath Airways passenger

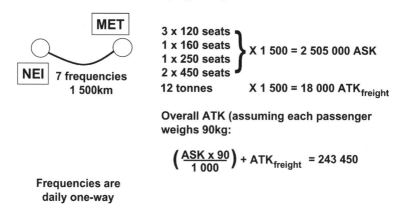

MET

NEI 7 frequencies
       1 500km

3 x 120 seats ⎫
1 x 160 seats ⎬ X 1 500 = 2 505 000 ASK
1 x 250 seats ⎪
2 x 450 seats ⎭

12 tonnes   X 1 500 = 18 000 ATK$_{freight}$

Overall ATK (assuming each passenger
weighs 90kg:

$$\left(\frac{ASK \times 90}{1\ 000}\right) + ATK_{freight} = 243\ 450$$

Frequencies are
daily one-way

Figure 3.4   Measuring Productivity – The Capacity

load factor for the Metropolis to Neighbourhood route is 73.8% and the overall, or weight, load factor for the route is 71.4%.

It is quite usual for overall load factors to be inferior to passenger load factors. This is because freight tends to be unidirectional, making it more difficult for airlines to use their capacity efficiently.

Target load factors should always be set. If average demand per flight is allowed to rise there will be some occasions when demand will exceed the capacity of the aircraft. The resulting demand spill, which we will examine in more detail later in this Chapter, represents lost revenue opportunity. If we already have an idea of how much spill is being generated, this will assist in fixing the target load factor.

When spill is low, then load factor can be allowed to rise; when spill is high, it is much better to keep load factors low in order to minimise the risk of turning away demand on some occasions. Load factor targets might differ according to the different market conditions on different sub-networks.

Balancing the Equation

Essentially, we are balancing three parameters: the capacity, traffic and load factor. Extrapolating any two of these will suggest the third. In order to establish a macro fleet requirement we must bring two elements into balance. To simplify the calculation we shall work with the passenger market, although the same principles apply to the overall payload. One of these two elements is the level of achieved sales, represented by the amount of RPKs generated throughout the network and the load factor. The second is the supply, represented by the number of aircraft at our disposal and their productivity in terms of ASKs. Knowledge of the RPKs and load factor enables us to compute the number of required ASKs. Figure 3.6 explains this relationship.

An alternative to working with sector distances and frequencies is to use the average speed of the aircraft and their annual utilisation. At a macro level, this can suffice. What we must achieve in our macro calculation is a close balance between the ASKs we believe we can produce, and the ASKs required to ensure that appropriate load factors are realised.

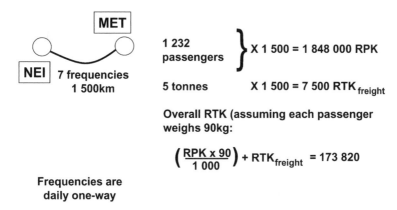

Figure 3.5   **Measuring Productivity – The Sales**

In order to fix numbers for the base year we must extract traffic data and the airline schedule. Then, a forecast must be considered of the two data sets. This forecast should take into consideration the evolution of demand, load factor and aircraft utilisation. The latter has enormous significance in economics and is dramatically affected by network shape, with long sectors enabling relatively high values due to the lower amount of unproductive time spent by aircraft on the ground. Airlines serving markets composed of short sectors tend to find that their fleet spends a higher proportion of time on the ground, so high utilisation is more difficult to achieve.

If we assume that load factor rises over time, then the number of ASKs required will fall, which means that less aircraft would be required. We shall see later in this Chapter that there is a limit to the extent to which we should see load factor climb due to the effects of demand spill.

If we assume that utilisation rises over time, then the number of ASKs produced will rise, which also means that less aircraft would be required. Although it is

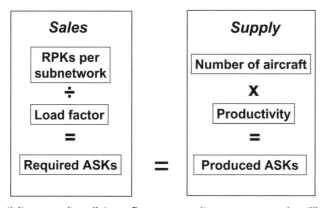

Productivity = capacity x distance flown = capacity x average speed x utilisation
or = capacity x sector distance x frequency

Figure 3.6   **The Business Equation**

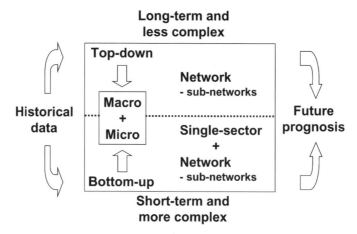

Figure 3.7   Two Approaches to Fleet Planning

obviously a goal to push utilisation as high as possible, sufficient margin should be built in for maintenance requirements and schedule recovery time.

Armed with only a handful of parameters we can now build our macro fleet plan.

A Macro Approach to Fleet Planning

Taking an overall view of what is going on has numerous advantages, the main ones being that limited resources and input are required to get an initial result. It is no exaggeration to say that many fleet plans involving enormous expenditure have been initially concocted on the back of an envelope. Macro planning involves aggregating data either for the entire network or, if the routes or markets served have different characteristics, a number of smaller sub-networks. Macro plans can be built quickly, are versatile in that the small number of basic assumptions can be rapidly altered, and require relatively little data and no complex analytical model to get a result.

Figure 3.7 illustrates the general architecture of the approach we will adopt. Whatever type of fleet plan is adopted, we should always draw from historical data, preferably for the entire market under study, rather than just for our airline. Problems obviously arise in fleet planning for start-up carriers where no history exists, or for route expansion, where data may be minimal. Once analysis has been completed for today's situation, we must then make a forecast of our needs. Therefore, we need inputs of future demand levels, as well as economic and market conditions.

Our macro fleet plan is divided into two parts, one comprising the demand (Table 3.1) and the second comprising the supply (Table 3.2). Each part of the plan is further sub-divided into several elements. The period under study is six years which, although rather short for a macro plan, is sufficient for the purposes of illustration. A typical macro plan could extend for double this period.

Table 3.1   Macro Study – The Demand

Expected traffic growth (%)

	2007	2008	2009	2010	2011	2012	2013
MET-NEI		1.05	1.03	1.03	1.03	1.03	1.03
Regional network		1.07	1.06	1.06	1.04	1.04	1.04

Annual RPK projection (millions)

	2007	2008	2009	2010	2011	2012	2013
MET-NEI	1345.0	1412.3	1454.6	1498.3	1543.2	1589.5	1637.2
Regional network	3833.0	4101.3	4347.4	4608.2	4792.6	4984.3	5183.6

Expected load factor evolution

	2007	2008	2009	2010	2011	2012	2013
MET-NEI	73.8%	75.0%	75.0%	75.0%	75.0%	75.0%	75.0%
Regional network	68.9%	70.0%	71.0%	72.0%	73.0%	74.0%	75.0%

Annual ASKs required (millions) = RPK / Load factor

	2007	2008	2009	2010	2011	2012	2013
MET-NEI	1822	1883	1939	1998	2058	2119	2183
Regional network	5563	5859	6123	6400	6565	6735	6912

Macro Study – The Demand

Expected traffic growth   If we believe that growth is likely to be significantly different within different parts of the network, then this alone is good reason for creating a number of sub-networks. In this case, our two sub-networks will grow at slightly different rates over the study period.

Annual RPK projection   In our base year, 2007, we have taken the actual RPKs recorded for Goliath Airways for the two sub-networks. The future values are derived from forecast growth rates. There are many and varied ways to forecast demand and these techniques are not the subject of this book. Suffice it to say that

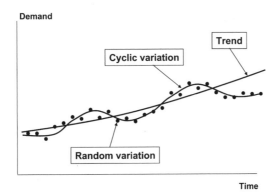

Figure 3.8   Demand Variability

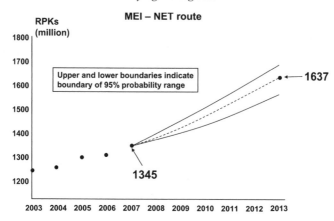

Figure 3.9   A Demand Forecast

it is important to bear in mind that future demand or traffic will vary according to (hopefully) predictable cyclic variations, as well as purely random variations that we shall explore in a moment. Also, there will be an underlying trend describing the overall evolution of the demand (Figure 3.8).

To help us build our macro fleet plan we need to take our base year of 2007 and apply, for each sub-network of the network, a forecast that describes a situation where we might expect a 50% probability of demand being either greater than or less than the line. It is also useful to describe two further curves that form the boundary of probability of the demand lying with (say) 95% of that range. Such a forecast can be achieved through several methods. For example, we might apply an econometric or else a behavioural model, or indeed we might select different models for different sub-networks. It is definitely important to work with a selection of scenarios as it is not statistically useful to base a fleet plan on a single set of figures. There is always uncertainty concerning the inputs to an econometric model. How easily can we measure future fares, Gross Domestic Product, personal disposable income, population, or any other explanatory variable?

Where the random, or stochastic, nature of demand should be simulated, the Monte Carlo technique can be a useful approach whereby random samples produced by a software with a random number generator are mixed as if they were fully independent.

Figure 3.9 shows a sample forecast of the MET to NEI route for our macro fleet plan.

Expected load factor evolution   The values from 2007 are those observed from historical data. Setting future load factor targets will dimension the amount of production that will be necessary in order to achieve the expected RPKs. Average load factors that are allowed to climb too high will result in demand spill, the relevance of which will be examined in our discussion on micro fleet planning later in this Chapter. We can assume that Goliath Airways would be happy to see a gradual improvement in their load factor up to a maximum of 75% for both sub-networks.

Table 3.2   Macro Study – The Supply

Annual ASKs produced (millions)
= number of aircraft x capacity x utilisation x average speed (714 kmph)

	2007	2008	2009	2010	2011	2012	2013
MET-NEI	1822	1884	1920	2009	2099	2142	2168
Regional network	5562	5853	6115	6403	6525	6740	6926

Expected utilisation evolution (per aircraft unit)

	2007	2008	2009	2010	2011	2012	2013
MET-NEI							
S120	1146	1160	1180	1180	1200	1100	1000
S160	191	220	240	300	400	400	400
S250	509	540	560	560	1000	1000	1000
S450	1528	1550	1550	1550	0	0	0
Regional network							
S120	1981	2000	2120	2120	2100	2160	2160
S160	2730	2900	3060	3000	2900	2900	2900
S250	2476	2600	2740	2740	2300	2300	2300
S450	1653	1750	1750	1750	0	0	0

Number of aircraft required

	2007	2008	2009	2010	2011	2012	2013
S120	4.0	4.0	4.0	4.0	6.0	7.0	8.0
S160	8.0	8.0	8.0	9.0	9.0	9.0	9.0
S250	3.0	3.0	3.0	3.0	6.0	6.0	6.0
S450	2.0	2.0	2.0	2.0	0.0	0.0	0.0

Annual ASKs required   This is the third of the three elements, that must be brought into balance. It is the result of the expected traffic and load factor.

Macro Study – The Supply

Annual ASKs produced   The value for 2007 is a computed value based upon the product of the number of aircraft in the fleet, their capacities, hours of utilisation and speed. The aim is to manipulate these inputs in order that the ASKs that would be produced are as close as possible to the ASKs *required*, which we have already determined in Table 3.1. If the numbers of aircraft in the fleet increase in order to accommodate traffic growth, then the utilisation of each unit must decline to some extent. The aircraft speed has been assumed to be a constant 714 kilometres/hour, although in reality we would need to input correct values according to each route and aircraft type.

Expected aircraft utilisation evolution   The operation is divided into two sub-networks as we presume that each will experience different growth levels and have different characteristics. We must likewise divide-up the aircraft utilisation between the sub-networks. The frequencies are known for 2007 and it is an easy matter to

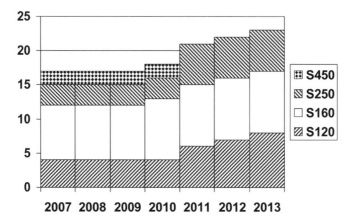

Figure 3.10   Goliath Airways Fleet Plan – Aircraft

work out the precise number of hours needed in each case. However, in macro fleet planning it is acceptable to grow the utilisation numbers without directly matching the values to actual flights. The purpose of the exercise is to determine an overall requirement based upon a limited set of inputs. We should not attempt to go too far and become embroiled in scheduling.

The maximum acceptable utilisation per aircraft type should be determined. In the example we shall assume this to be 3,300 hours. This is a reasonable target given the stage lengths of our operation. The values for 2007 are actual numbers from the existing operation, whereas from 2008 the numbers are estimated. Our utilisation target means that there is still some slack in the operation as of today so, given the traffic growth and target load factors, the required ASKs can be achieved without increasing the fleet size until 2010.

Number of aircraft required   From 2010 onward, the only way that sufficient ASKs can be produced is to increase the number of aircraft, so an additional S160 unit is taken that year. However, Goliath Airways is still operating four aircraft types for a relatively small operation and there is considerable disparity between them. We shall see in Chapter 6 that there are economic implications of poorly spacing aircraft sizes in a fleet. Let's suppose that Goliath would like to take the opportunity of disposing of its largest capacity type, the S450, of which it is only operating two aircraft. In 2011, these two units are presumed sold, but sufficient capacity must, of course, be brought in to ensure that enough ASKs can be produced. Thus, the S250 and S120 fleets are now grown. Every time a capacity change is made, a new balance has to be found with expected traffic, load factor and utilisation.

Figures 3.10 and 3.11 reveal the progression of the number of aircraft and the number of seats in the fleet throughout the plan.

Over the six years of the study the number of aircraft in the fleet has grown from 17 to 23, and the number of types reduced from four to three. Such streamlining is all very well, but we could have achieved the same matching of 'ASKs required' and 'ASKs produced' by many other combinations of input. A major drawback of macro fleet planning is that it is impossible to assess the effect of any change in either the traffic or operating patterns.

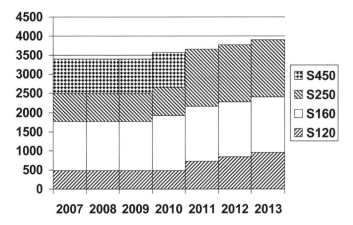

Figure 3.11   Goliath Airways Fleet Plan – Seats

A macro plan cannot easily differentiate between four S120 (120-seater) aircraft, or two S250 (250-seater) aircraft. A market deserves one solution or the other.

As with any planning activity, the results you get are only as good as the assumptions you work with. Any change to the growth rates or the achieved load factors would seriously undermine the capacity requirements. Also, no account has been taken in the plan of the forthcoming appearance in the market of the start-up Barracuda Airlines. An incumbent airline would be well advised to reflect the appearance of a new competitor but, given the track record of some start-ups, Goliath Airways might be forgiven for not rewriting its fleet plan completely until Barracuda has become established.

Macro fleet planning provides only the broadest estimate of what is going on and is considered by many to be too simplistic and lacking in sophistication. Nevertheless, the approach we have just outlined is an excellent starting-point to get a grip on the magnitude of the capacity needs. A critic will point out that a macro plan guarantees that you know nothing about everything.

A Micro Approach to Fleet Planning

Taking a look at a network from the perspective of individual routes, city-pairs and flight numbers gives us an opportunity of accurately reflecting the real operation. If we can concoct a model that allocates demand, firstly according to a pre-determined set of criteria, and secondly to each segment on a network by flight number, then we have an opportunity of calibrating the prediction with historical values.

It is far from easy to build such a model, as we must compress all the complexities of demand behaviour to a small number of criteria that determine how individual flights are filled. Yet, if we can succeed in replicating an operation with an appropriate statistical confidence level, then a demand allocation program is extremely powerful. In fact, once such a model has been built it can be easily used to test different capacity solutions, such as changing aircraft sizes. When linked to an economic calculation module, we have the makings of a fully integrated fleet planning tool.

However, there are drawbacks to micro planning. Firstly, demand allocation models are only truly valid for short-term projections. Market conditions change rapidly, with route additions and deletions, new aircraft types and competitors emerging and fading. The past does not guarantee the future. If we base our model purely upon observed trends and data, we might just as well start walking backwards down the street. If we truly believe that the pavement will continue in the same direction, sooner or later we will bump into something. This problem tends to limit complex models to relatively short-term use, say, up to three years.

A second problem with complex micro models is that they require large amounts of accurate data. Many airlines do not organise data on a true origin and destination basis. Segment, or sector, data are more readily available. Indeed, complex networks can accommodate many hundreds of real origin and destination data sets, and it is time-consuming to track these down unless we have easy access to MIDT data, for example.

Our third problem is that, owing to the large data sets required, plus the time, knowledge and experience levels necessary to process these data, we must accept that this planning approach is resource-heavy. Getting useful results takes time, yet senior management often wants quick answers! Micro plans become hugely dependent on the quality of the inputs and it is all too easy to get lost in a morass of detail. We have seen that a macro fleet plan means you end up knowing nothing about everything. With a micro plan you know everything about nothing.

To make life even more complicated it is essential to predict future levels of the elements that make up any form of model, whether they are demand levels, economic and financial inputs and so on.

Macro models are relatively straightforward to construct, whereas micro models based on a complete network tend to be limited to airlines with sufficient planning resources, manufacturers and specialists.

Whatever kind of modelling technique is adopted, there is a single starting-point, which is building an understanding of the market and its behaviour. It is to this central element that we now turn.

Market Segmentation and Spill

A successful fleet plan should take account of demand behaviour and map as far as possible the different components, or segments, of the demand. The first part of this section will address travel motivation. We then move into an analysis of demand spill and examine how to treat variations of demand over time in the fleet plan. Although the discussion that follows centres on passenger demand, many of the principles apply to the cargo market.

Motivations of Travel Demand

Segmentation of markets is not a new technique and is practised widely in the airline industry and beyond. There is a tendency for segmentation to become over-complex, particularly as revenue management systems have encouraged the creation of a large number of booking classes. For example, British Airways operate 26 booking classes, each of which has the potential to be sub-divided into 10 further portions. Clearly, segmentation of this order of magnitude has value in revenue management terms, but is beyond the scope of fleet planning because it is difficult to model and is intended to help maximise profits rather than help choose an aircraft.

For airlines that operate a multi-class aircraft cabin it is essential to decide whether it is better to segment the travel market by physical class of the aircraft, or else according to other criteria. Segmenting by aircraft class has the advantage of simplicity, although it is becoming more and more apparent that a physical class within the aircraft may comprise passengers travelling for diverse reasons. For example, North American business travellers tend to select Coach-class, whereas European business travellers tend to favour Business-class. There are many indicators that this particular trend is evolving, with more and more cost-conscious company travel departments insisting that employees migrate towards the back of the aircraft. Some airlines, which were initially regarded as catering to the leisure market, such as easyJet, have actually become attractive to business travellers.

In order to arrive at an appropriate segmentation, the airline should examine its complete market, identify specific traits, and group these into a reasonable number of categories. For carriers with a route network embracing several geographical regions it may be necessary to define several segmentation categories. It should also be recognised that elements seemingly important to a short-haul passenger are not at all important for a long-haul passenger.

In compiling market segmentation for fleet planning purposes, we are aiming to identify the magnitude of each category, its growth potential, the balance between revenue potential and cost of service provision and the sensitivity of the demand

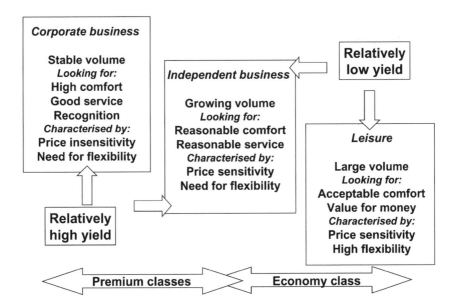

Figure 3.12 Market Segment Characteristics

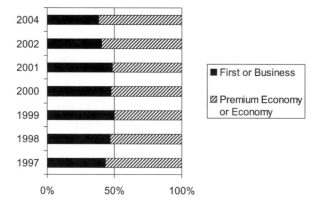

Figure 3.13   Usual Class of Long-Haul Travel
Source: Original data IATA

to changes in the schedule. Each market segment comprises a different booking behaviour, knowledge of which is essential in order to economically size both the aircraft and each cabin.

Life becomes complicated because there are occasions on which so-called 'business' passengers exhibit price insensitive behaviour, and likewise, 'leisure' passengers may equally exhibit price sensitive behaviour. We must also bear in mind that the traditional business market is fragmenting into two categories. Firstly, we have the corporate business market, which includes people working for (guess what) large corporations, plus governments and any other large institution. Then we have a growing sector populated by business people who hold a greater personal responsibility either for the company they are working for, or else the travel budget. This is the sort of market that gravitates to the low-cost carriers.

The leisure market could be sub-divided into a number of areas. We may need to distinguish between families, individual adventurers, the 'silver-haired' market, the young student market, the visiting friends and relatives market, ethnic travellers, educational travellers and so on.

Indeed, the low-cost carriers may claim credit for the emergence of a wholly new breed of week-end traveller, who perceives that the price of an air ticket is hardly more than that of a meal in a restaurant.

Even when travelling for leisure, those passengers who may have been looking for a great deal may also be prepared to pay a premium for extra leg room. Indeed, Monarch Scheduled has removed one seat row in their A320 and A321 aircraft in order to create 34 inch extra seat pitch in seven rows. They found that passengers are quite happy to pay a premium for the extra pitch that is available in the 42 available seats on these aircraft types.

In constructing market segments, it may be found that some people feel a loyalty to their national carrier, for example. Others may simply be averse to taking what they perceive as a risk of flying with a less familiar airline. Another category may be motivated purely by frequent flyer miles, and is therefore likely to select airlines adhering to one of the major alliances. Yet another might adopt the more pragmatic

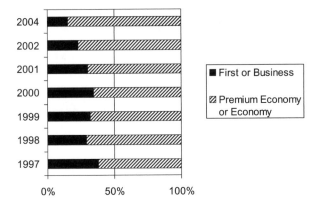

Figure 3.14 Usual Class of Short-Haul Travel
Source: Original data IATA

approach of selecting an airline purely according to a set of rational criteria, such as flight time and price.

All this makes market segmentation an obviously complex issue. The problem for the fleet planner is how to capture the essence of market segmentation in the context of the aircraft size required and, with some difficulty, in the dimensions of the individual cabins.

The IATA surveys The International Air Transport Association (IATA) undertakes a regular survey of several thousand passengers, who are travelling for business purposes, entitled the Corporate Air Travel Survey. The survey, which is conducted in North America, Europe and Asia-Pacific, provides a wealth of valuable information

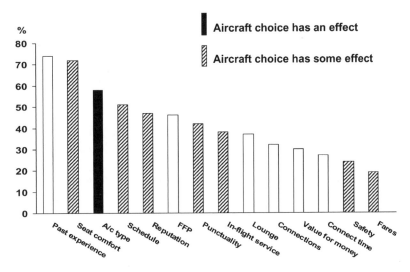

Figure 3.15 Factors Influencing Airline Choice, Long-Haul, First-Class
Source: Original data IATA

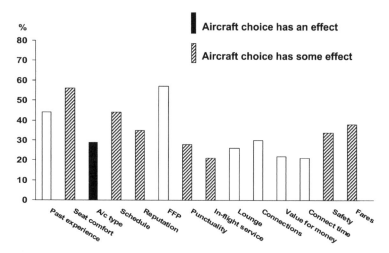

Figure 3.16 Factors Influencing Airline Choice, Long-Haul, Business-Class
Source: Original data IATA

about how markets are evolving and how passengers rate certain aspects of their travel experience. These data have relevance for fleet planners because in a competitive situation an airline's fleet choice is partly responsible for market behaviour.

Figures 3.13 and 3.14 show the degree to which the selection of class of travel has altered over the period 1997–2004 (there were no survey results for 2003). It can be observed that in both short and long-haul markets there is a gradual decline in the proportion of travellers in the premium classes. The justification for the elimination of First-class cabins in many airlines can be found in analyses of this type. Aircraft configuration decisions can be aided by an analysis of the target market.

The IATA surveys have also thrown light on the most important elements in the airline selection process for travellers in different classes. Figures 3.15 to 3.17 show these results with a consistent ranking of factors. It is rather unsurprising to note that passengers travelling in the Economy-class cabin have a very different set of decision criteria to those in First or Business-class. Note that Figure 3.17 combines the results of both Economy as well as Premium Economy-class travellers, where the differences were relatively minor.

The bars have been coded (by me and not by IATA) according to those elements where the aircraft choice might either have a direct or some effect.

First-class, and to some degree Business-class, passengers assign a considerable importance to seating comfort on aircraft compared to Economy-class corporate travellers. Such a result sends a clear signal to airline product managers as well as highlighting the importance of the design of the premium cabin.

Somewhat surprisingly, the actual aircraft type is a very high decision factor for First-class travellers, although this is much less the case for the other classes. Nevertheless, for airlines that have retained this high-yielding, albeit expensive, cabin it reveals the degree to which regular travellers can get to know the various aircraft types.

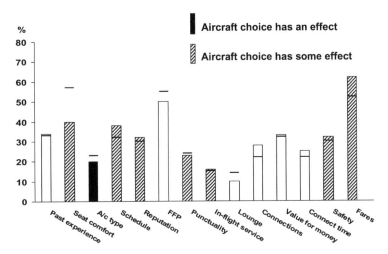

**Figure 3.17 Factors Influencing Airline Choice, Long-Haul, Economy and
Premium Economy-Class**

Source: Original data IATA

The availability of low fares is, again unsurprisingly, of less importance for First-class passengers who, in the main, are not paying for their own tickets. This latter point is borne out when we compare the degree to which company travel policy affects choice between the three classes of travel.

The availability of low fares is relevant to aircraft choice insofar as an airline only has the ability to compete vigorously on price if it is able to operate a cost-effective aircraft. Profit margins are very thin in the airline business, and a successful low-fare policy is highly dependent upon the low operating economics of the aircraft.

Variations in Travel Demand Over Time

The next component in the jigsaw is to take a closer look at the degree to which the demand will vary over time. There will be enormous differences according to the length of the route and the type of traffic carried.

Here we need to make the distinction between demand and traffic. We may be quite good at forecasting demand, but the traffic is the result of how much of that demand we are able to carry as a function of being able to provide the right amount of capacity at the right time, measured over a period of time. There will be some occasions when demand will exceed the capacity supplied. This can be illustrated in Figure 3.18, which describes a situation of demand variability by day of the week.

Business-oriented short-haul routes having around a ninety-minute flight time often have demand peaks in the morning and evening. Routes that fall into this category include Paris–Toulouse, Johannesburg–Cape Town and Melbourne–Sydney. In each of these cases, there is a pronounced and somewhat compressed peak in the morning, a mini-peak in the middle of the day, and a somewhat elongated evening peak that does not quite reach the magnitude of the morning peak.

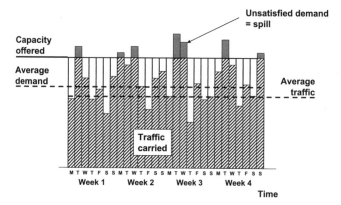

Figure 3.18 How Spill is Generated

The timing of the peaks is almost the same in both directions. An immediate problem, which is faced on such routes, is that a large aircraft scheduled to coincide with one of the peaks in one direction must return in the opposite direction when the peak has passed. Clearly this presents a problem. Satisfying peak demand is only possible at the expense of seeing the aircraft fly in the opposite direction at a lower load factor than the demand would warrant.

An important parameter to establish, therefore, is the peak-to-low ratio, or else the ratio of the 'peakiest' period of time to the lowest. In Figure 3.19, we might wish to consider separately the ratio of the peak to the trough of demand in the morning period as well as that of the middle of the day. The much lower demand in the 06:00 to 08:00 time period would clearly warrant a smaller aircraft, whereas the less severe depression of demand in the middle of the day might still be dealt with by the same size of aircraft serving the peaks, although at a penalty. A trade-off has to be calculated between minimising the number of aircraft types on the route and a better matching of capacity to the dramatic variation in demand.

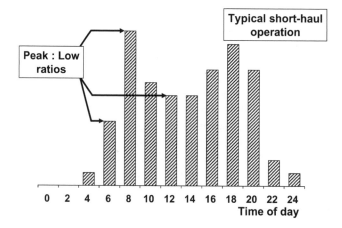

Figure 3.19 Daily Demand Variation

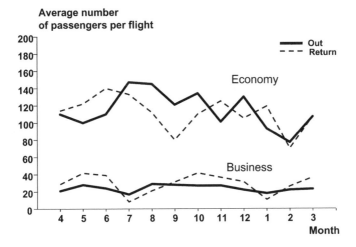

Figure 3.20 Seasonal Demand Variation

If we expand the time period to a week, we might easily find a similar problem. Weekly demand might, for instance, peak at the beginning and end of a week, when business travel is at its height. Further expansion into a season will doubtless reveal even more variations in demand magnitude. In the example in Figure 3.20, we can see that there is a dramatic variation of Economy-class demand between July and September, according to direction travelled. Business demand is also quite unstable, both by month of the year and by direction. The problem is how to plan the capacity to serve these two markets economically. In order to do this we need to apply statistical techniques to measure the variation of demand around the mean value.

Modelling Spill for Fleet Plans

Spill is the degree of average demand which, for a particular time period, exceeds the capacity offered. Simply put, spill is lost demand. You may well consider that providing insufficient capacity for the market can only be a bad thing, and that airlines should do all they can to accommodate demand. Hopefully, after reading this analysis you will conclude that spill is actually a *necessary* evil, even though regularly incurring spill can damage the business in the longer term.

We have already considered that markets divide up into a number of individual segments, each with an identifiable behaviour. In order to appreciate the effect of spill, we shall start by assuming a unique market segment, defined as a grouping of passengers exhibiting similar characteristics in terms of need and expectation, and then move into looking at how we might consider a multitude of segments.

From historical data we may see that over a period of time there were a certain number of occasions when a particular flight was full, some occasions when there was spare capacity, and some occasions when the demand exceeded the capacity. Over a particular time period we can therefore determine the average load. This can be compared to the average capacity, being a measure of the total number of flights

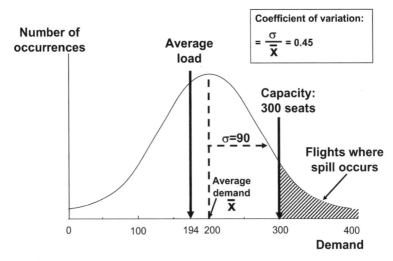

Figure 3.21 Modelling the Spill (1)

provided divided by the total number of flights operated. The ratio of average load and average capacity is the average load factor.

If the airline could have accommodated every single passenger who wanted to travel on the flight we could establish the unconstrained demand. We never actually know the precise value of this, because there will always be occasions when an attempted reservation has been thwarted by, for example, inability to contact a reservations office, access to the airline website, or else the emergence of an alternative offer.

The Normal Distribution

To simplify matters, fleet planners have historically made the presumption that demand behaves according to a probabilistic law. So, for convenience we use the normal, or Gaussian distribution. Thus, the mean value of the normal distribution represents the unconstrained demand.

In any situation where there will be *some* occasions when demand exceeds available capacity, this means that the average number of passengers carried will be slightly inferior to the average demand, as we saw in Figure 3.18.

The shape of the distribution determines the spread of demand, which is measured by the standard deviation, represented by the Greek symbol sigma. By convention, we work with a coefficient, rather than absolute values, in order to have a standard parameter to describe the different market segments, and to ease comparisons with markets of different sizes. This coefficient is the ratio of one standard deviation to the mean. We call this the coefficient of variation, or else the k-factor.

In the example shown in Figure 3.21, the average demand for the market is 200 and one standard deviation is 90 passengers. The law of the normal distribution states that 68% of all observations are to be found within plus or minus one standard deviation from the mean, or within the range 110–290. If this were not to be the

Table 3.3   Spill Calculation

K factor	Capacity	Average demand	Average load factor	Average load	Spill
0.45	300	160	53.1%	159.3	0.7
		180	59.2%	177.5	2.5
		200	64.7%	194.0	6.0
		220	69.4%	208.3	11.7
		240	73.5%	220.4	19.6
		260	76.9%	230.6	29.4
		280	79.7%	239.1	40.9
		300	82.0%	246.1	53.9
		320	84.0%	252.0	68.0

case, then the demand could not be normally distributed. With these parameters the k-factor is 0.45. If the capacity were set at 300 seats, then the curve suggests the amount of demand that could not be accommodated on an average basis. The larger the k-factor, the broader becomes the distribution of demand around the mean, which may, of course, stay the same.

The average number of passengers carried will be slightly inferior to the original demand, the difference being explained by the area to the right of the aircraft capacity. Table 3.3 shows the calculation of average loads for a range of demand values. It can be seen that as average demand rises then spill rises rapidly. The calculation specific to the average demand depicted in Figure 3.21 is highlighted in the table. Thus, six passengers would be spilled on an average basis, resulting in an expected average load factor of 64.7%.

In Figure 3.22 Case A represents a market with an average level of demand of 70 passengers. However, depending upon the distribution of demand around the mean, the degree of spill generated will be quite different. If the market does not exhibit

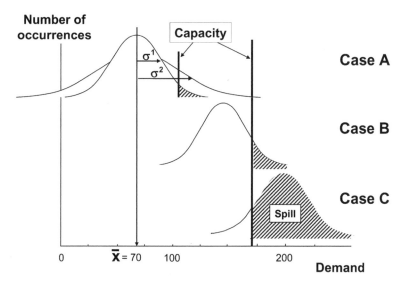

Figure 3.22 Modelling the Spill (2)

wide variations the standard deviation will be relatively low. In the example the value is 17 (represented by δ^1). On the other hand, if the demand is widely dispersed the standard deviation will be higher, in this case 50 (represented by δ^2). The two k-factors are calculated as follows:

$$\delta^1/\overline{x} = 0.24$$
$$\delta^2/\overline{x} = 0.71$$

Imagine that we fix the capacity at 100 seats. For the case where the k-factor is 0.24, the average spill generated would be less than that generated for the higher k-factor. So, in order to reduce the spill for the higher k-factor case, we would need to increase the capacity.

Mapping the demand distributions by market segment enables us to measure the degree of spill likely to be generated as demand increases. In Figure 3.22 the demand can be presumed to increase over time. Therefore, we can see at a glance the effect that capacity changes might have on the amount of demand directly satisfied. If the *shape* of the demand curve, and therefore the standard deviation, does not change, then as average demand expands the k-factor tends to reduce. This suggests that a closer relationship can be built between the mean demand and the aircraft capacity. However, there is a tendency for the standard deviation to expand as market size increases.

It is even possible for the mean demand to exceed the capacity of the aircraft. In Case C the mean is 200 whereas the capacity is only 170. Yet this does not mean that the aircraft will fully load on every occasion. There will still be some occasions when the aircraft will attract less than a full load. This explains why load factors rarely reach 100% on a permanent basis.

To summarise, the different demand curves that we have examined can represent different segments of the same market, with consequently different behaviour. So, we use the k-factor to determine the proportion of seats likely to be filled either for the whole aircraft or for a cabin within the aircraft.

The relationship between mean demand and capacity becomes especially significant because we know from experience that different market segments tend to exhibit different k-factors. Sometimes, business markets have a tendency towards high k-factors, and a consequently wide spread of demand, whereas leisure markets may have a tendency towards low k-factors, where demand is more tightly peaked. There are three basic reasons for this. Firstly, business demand may be irregular in origin, and thus difficult to predict. However, the occurrence of a conference, festival or other such activity would clearly stimulate demand. Secondly, business demand books very late, when capacity adjustment becomes very difficult. Thirdly, airlines prefer not to discourage business demand from particular flights by restrictions. This last point is important because we know that business travellers require and expect access to seats when they wish. Denial of access to seats is the fastest way to lose the loyalty of customers who contribute the greatest yield.

By combining our knowledge of market segment behaviour, booking preferences and the distribution of the various demand curves, we now have an opportunity to determine the best size of each cabin within the aircraft, according to the mean value of the demand, measured over an appropriate period. Also, now that we can estimate the amount of spill that may be generated for a given demand, we can determine the most appropriate cabin planning load factor. Once this cabin planning load factor is approached, or breached, then thought should be given to changing the configuration of the aircraft, or the numbers of frequencies operated, or even the aircraft type.

Dealing with negative demand Where mean demand is relatively low, and where the distribution around the mean is rather flat, indicating a high k-factor, we may find the left tail of the curve overlapping the point of zero demand. Mathematically, this is correct, so long as we continue to presume that demand is normally distributed. However, it is clear that negative demand cannot exist. There are two ways of dealing with this phenomenon. Either the negative demand can be redistributed under the area of the curve where demand is positive, or else it can be ignored. In fact, because the area under the Gaussian curve where negative demand occurs is rather small, we usually ignore the effect altogether.

Choosing the right time period to analyse The actual amount of spill is partly a function of the time period under analysis. For example, if a market is seasonal, with a strong tendency to peak during the summer months, then measuring demand and applying capacity on an annual basis would give a completely different result than if the measurement of demand were to be based on two seasons. In that case, varying the degree of capacity according to known seasonal demand variations would control the number of occasions when demand could not be satisfied. Naturally, it is not always practically possible to easily vary the capacity over a period of time. This depends upon the degree to which an airline would be able to redeploy aircraft on other parts of the network or else lease-in or lease-out units of capacity. Small operators with a limited number of routes may be constrained in this respect.

An example of k-factor analysis In Figure 3.23 we see an example of how an airline might set about establishing its own k-factor for, in this case, a particular flight. This airline conducted a survey of Business-class demand over a complete season for one flight. The capacity of the Business-class cabin was 40 seats. The survey picked up those occasions when demand exceeded the capacity of the cabin. It can be seen that on a very few number of occasions the demand for the flight was around 60, and on two occasions the demand was even non-existent! We can easily observe that the reported data do in fact reveal the makings of a normal distribution, from which we can calculate the mean, standard deviation and, therefore, the k-factor.

The objective of this particular exercise was to determine the k-factor in order to decide on the optimum size of the Business-class cabin. The only way to be sure of accommodating every single person wishing to travel would be to create a cabin with 62 seats. Economically this makes no sense, of course. As long as we can assess the value of each unit of demand, especially those who book late and risk being spilled, then an economic optimum can be found. We are balancing the cost of providing capacity that will be only used on an irregular basis with the loss of revenue on those occasions when spill is generated.

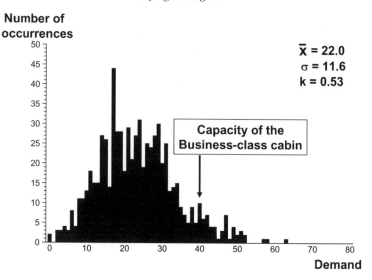

Figure 3.23 Business Demand Variability

So, yes, spill is a good thing, and this is particularly the case if you are a monopolist. Before deregulation came to France, the monopolistic predecessor of Air France was far less concerned about spill than a competitive player would have been. Despite the high-speed rail network many domestic routes in France did not have realistic competition from rail and there was no alternative for a spilled passenger than to accept a second-best flight. With the more competitive nature of many routes this situation has now changed and Air France is less complacent about its position.

Once the normal distribution curves have been calculated for each market segment, the next step is to express the curves as a relationship between the percentage of original demand spilled, and the load factor achieved. Figure 3.24 shows how increasing load factor rapidly leads to demand being spilled. Once the k-factor is known it is straightforward to determine an acceptable cabin load factor target.

To illustrate how these curves may be used, refer back to the spill calculation in Table 3.3. You will see that for an average load factor of 79.7% that the spill would be 40.9 passengers, or 14.6% of the demand of 280 passengers. Now if you refer to Figure 3.24 you can interpolate where the $k = 0.45$ curve would be and, at a load factor of just under 80%, simply read off the identical value of 14.6 on the y-axis.

A business-orientated market that has a wide dispersion of demand around the mean could be expected to have a relatively high k-factor, meaning that low load factors should be targeted in order to reduce the risk of spill. Conversely, a market with a much tighter distribution of demand around the mean is more forgiving, in that average load factor can be increased to higher levels before spill takes hold.

Critics of the use of this methodology may say that the normal distribution may not always be the most appropriate statistical measure of demand. When working with small values it is sometimes difficult to place one's hand on one's heart and declare that demand really does follow the Gaussian distribution curve.

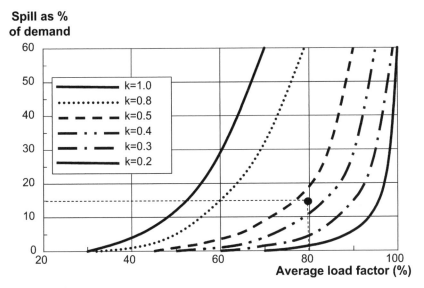

Figure 3.24 Load Factor vs. Spill

Another trap is that there is a tendency to assume that demand occurs when there is a service offered. In other words, the mere *presence* of a flight tends to attract demand. In order to ensure that we are making the correct capacity decisions, it is vital to determine the underlying demand curve shape, before any 'contamination', such as the schedule, takes effect.

For better or worse, this is the mechanism that a great many fleet planners adopt. The concept of spill has become integral to many planning and fleet assignment models, as well as revenue management programs. We shall now examine two uses of spill in fleet planning: firstly in a model to determine market behaviour on a single route; and secondly in a network model. In both cases the intention is to decide on the best mixture of aircraft size and frequency, for today and the future.

Modelling on a Single Sector

There are powerful arguments against modelling a single sector in isolation. Firstly, aircraft are rarely assigned on the basis of a particular sector or route. Exceptions are where an aircraft may possess unique performance or characteristics, or where a specific cabin configuration is required. For example, Singapore Airlines assigns its five A340-500 aircraft on purely the Singapore to New York and Los Angeles routes. This is due to the unique range requirements and special configurations needed for the market.

In most cases, aircraft are assigned to a multitude of routes making it inappropriate to place too much judgement on an aircraft's economic performance on a single route. Yet there will be occasions when a single sector might predominate to such an extent that the bulk of analysis could be directed to getting that part of the network right. The London–Dublin route is of paramount importance in the European network of Aer Lingus. Similarly, the challenge for Qantas is to find the optimum mix of capacity and frequency for the Melbourne–Sydney route. Once the prime route is optimised the rest of the network can follow.

Where a single sector is of such significance within a network that it warrants individual attention, then we can go ahead and apply specific modelling techniques to that sector alone. However, when analysing a single sector it is important to have a feel for whether the demand is true 'origin and destination' associated with the city-pair or whether there is contamination of the underlying demand travelling to and from other points in the network. In that case, we would need to consider whether or not it is the wider network that is driving the principal city-pair.

In the example, we shall take the high-density business-driven city-pair from our study network, Metropolis and Neighbourhood. The two airlines currently serving the route, Goliath Airways and Air Hercules, are about to be joined by Barracuda Airlines, a start-up carrier.

At this stage, we shall make broad estimates of the trip cost of the aircraft flying on the route, as well as the yield generated. As the evaluation progresses and aircraft types being considered by Barracuda Airlines become more closely defined it would be quite usual to return to the calculation later to fine-tune the inputs. We shall also assume that Metropolis Airport is slot constrained and that demand is highly sensitive to frequency.

A number of parameters are already presumed to be available. For example, the schedule of the incumbent airlines is public knowledge and the operating costs of their aircraft can be reasonably assessed. Also, it is usually possible to establish the final load factors being currently achieved as well as make a fair estimate of the average yield based upon ticket prices. It is virtually impossible to assess a competitor's true revenues, as this is highly dependent upon commission levels, discounting, and the degree of pro-rating. However, with significant transparency of pricing thanks to the Internet, it is certainly possible to arrive at a fair estimate. One problem a start-up might face is that it would be unlikely to afford to purchase MIDT information, which would give an accurate picture of booking levels. However, let's assume that the study is being performed on behalf of Goliath Airways, who are fortunate in having access to such data.

It is usually advisable to analyse a single-sector in both directions, as demand peaks may not coincide. This is often the case for routes feeding large centres of population or hubs. Thus, the morning demand peak for the Toulouse–Paris route occurs slightly earlier than the corresponding peak in the opposite direction because it takes longer for passengers to travel from Orly airport in Paris to the centres of business than for Parisians to make the much shorter journey to the centre of Toulouse. The corresponding evening peaks are likewise staggered in each direction. Similarly, FlyBe realised that their London to Belfast travellers were looking for a long working day in London, but a shorter working day in Belfast. It made sense, therefore, to run the operation from Belfast rather than London so the schedule could be more easily adapted to the demand.

When we look at how demand is spread over the week we might find that there is a cluster of high demand on Mondays and Fridays. This needs to be identified in case aircraft substitutions would be required. It is more usually the case, however, to adopt data from a single day. Amassing sufficient data from a series of mid-week days would suffice.

Before we start to process any data we need to remember that demand does not equal traffic. As we have already seen, spill is that part of the demand that has not been carried, so whenever spill occurs, demand and traffic cannot be the same thing. Although we can never actually know the original demand we can make a fair guess. There are two ways of doing this. Firstly, so long as we already have some idea of what the k-factor might be, we could make a spill calculation in reverse – so-called 'de-spill'. Secondly, if we can access booking curves for each flight we might have a better understanding of how the shape of demand builds for the total market before any spill has actually occurred. Thus, we would be taking the booking data at some point before each flight has departed as a proxy for the demand shape for the market.

We now have enough elements in hand to make a preliminary estimate of how the market might behave. To summarise, we have:

- *Fleet details* (aircraft types normally assigned to each flight number)
- *Timetable* (for a given average day, by direction)
- *On-board loads* (averaged for the period, to be used for calibration)
- *Booking curves* (to give an indication of the demand shape).

Calibrating the Model

Before processing the data we need to construct a mechanism that allows us to track variations of demand according to time of day. For the purposes of this illustration we shall divide the operating day into segments of 15 minutes. These segments must be small enough so that each can accommodate no more than one flight in the schedule.

The analysis we shall make will consist of constructing a demand curve shape so that the area under the curve equates to the total expected demand for the chosen day of study. The smooth demand curve means that each 15-minute time period we have adopted in the model will comprise a certain amount of demand. This demand is then compared to whatever capacity is offered at that time. For every 15-minute time slot with no capacity, all of the demand will logically be 'spilled'. When capacity

Table 3.4   Metropolis to Neighbourhood Pre Deregulation Timetable

Time		Flight number	Aircraft type	Average fare ($)	Average recorded load
06:50	Goliath Airways	0011	S120	170	112
07:30	Air Hercules	0201	S300	180	250
08:35	Goliath Airways	0012	S450	190	260
09:20	Air Hercules	0202	S160	180	125
10:30	Goliath Airways	0013	S160	165	85
11:30	Air Hercules	0203	S120	165	90
12:00	Goliath Airways	0014	S120	180	80
13:15	Air Hercules	0204	S160	180	120
15:30	Goliath Airways	0015	S120	165	115
17:20	Air Hercules	0205	S250	190	200
19:15	Goliath Airways	0016	S450	190	390
19:55	Air Hercules	0206	S300	180	275
21:00	Goliath Airways	0017	S250	160	190
21:55	Air Hercules	0207	S160	160	127

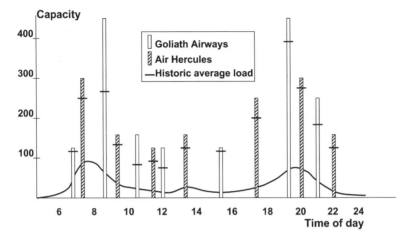

Figure 3.25 Metropolis to Neighbourhood Demand Curve

is offered, the demand will be compared and spill calculated according to a pre-determined k-factor. In this case the k-factor is presumed to be 0.4.

A reallocation process directs the entire spill to adjacent flights. For example, spill generated in the morning could be channelled to a preceding flight, whereas spill in the evening could find later flights. The reallocation process can be determined by examining real demand behaviour. The model can work in a series of loops, so that successive runs are comparing demand and capacity in each 15-minute time slot, with reallocation of the spill going on between each run.

Table 3.4 summarises one direction on the study route with the flight numbers of all aircraft operated, average fare per flight, and average load. The initial intention of the model is to replicate as closely as possible the recorded loads.

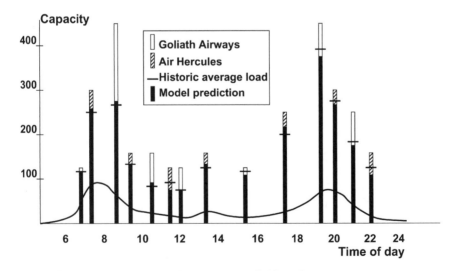

Figure 3.26 Metropolis to Neighbourhood Calibration

Table 3.5 Metropolis to Neighbourhood Post Deregulation Timetable

Time		Flight number	Aircraft type	Average fare ($)	Average recorded load
06:50	Goliath Airways	0011	S120	170	112
07:30	Air Hercules	0201	S300	180	250
07:50	Barracuda Airlines	4001	S160	140	
08:35	Goliath Airways	0012	S450	190	260
09:20	Air Hercules	0202	S160	180	125
10:30	Goliath Airways	0013	S160	165	85
11:30	Air Hercules	0203	S120	165	90
12:00	Goliath Airways	0014	S120	180	80
13:15	Air Hercules	0204	S160	180	120
14:00	Barracuda Airlines	4002	S160	120	
15:30	Goliath Airways	0015	S120	165	115
17:00	Barracuda Airlines	4003	S160	140	
17:20	Air Hercules	0205	S250	190	200
19:15	Goliath Airways	0016	S450	190	390
19:55	Air Hercules	0206	S300	180	275
20:30	Barracuda Airlines	4004	S160	170	
21:00	Goliath Airways	0017	S250	160	190
21:55	Air Hercules	0207	S160	160	127

Figure 3.25 shows the demand curve shape, with the Goliath Airways and Air Hercules flights plotted within the appropriate 15-minute segment along the *x*-axis. The small horizontal bars on each flight represent the recorded loads.

Once a demand curve has been estimated a number of iterations is necessary to ensure that the parameters we have entered into the model reproduce as closely as possible historical data from the two incumbent airlines. Thus, the k-factor, spill reallocation model and demand curve are modified so that, when the timetable is applied, all the flights operated by the two airlines are seen to generate loads that are close to those actually experienced. A statistical test is usually conducted to ensure that predicted and actual loads are within (say) 95% confidence limits.

Figure 3.26 shows the results of a series of iterations of the model. The simulated loads can be observed to be fairly close to the horizontal bars representing actual traffic.

When the calibration has been completed it is then possible to freeze the parameters which allocated demand and modify the schedule to reflect the deregulated market and the addition of the newcomer, Barracuda Airlines.

Barracuda Airlines flights must now be added. We have assumed that the newcomer would mount four frequencies with a single aircraft type, the S160, and that they would have the ability to schedule their departures very close to the two demand peaks. Table 3.5 shows the revised timetable.

Table 3.6 Metropolis to Neighbourhood Market Simulation

Before deregulation

	Revenue ($)	Cost ($)	Frequencies	Market share	Load factor	Spill
Goliath Airways	220748	79535	7	51.0%	73.8%	148
Air Hercules	211952	75435	7	49.0%	81.6%	153
Total	432700	154970	14	100.0%		301

After deregulation - no reaction of incumbents

	Revenue ($)	Cost ($)	Frequencies	Market share	Load factor	Spill
Goliath Airways	179868	79535	7	41.4%	59.7%	99
Air Hercules	161148	75435	7	37.3%	62.1%	70
Barracuda Airlines	73669	32656	4	21.3%	80.3%	98
Total	414685	187626	18	100.0%		267

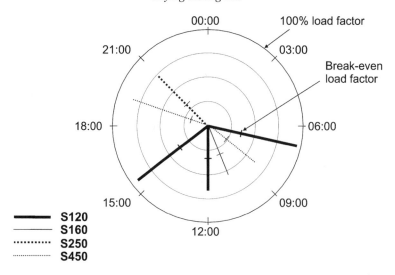

Figure 3.27 Goliath Airways Before Deregulation

By re-running the model at this stage we can begin to judge the damage that may be caused to Goliath and Hercules. What has happened in the simulation is shown in Table 3.6.

Before deregulation the two incumbents had divided the market almost equally between each other and were achieving reasonable load factors. Goliath Airways' load factor of 73.9% was inferior to that of the competitor because they were scheduling their very large S450 aircraft on the route. The total amount of spill generated, 301 passengers, was completely recaptured so the total amount of passengers carried, in the order of 2,400, equates to the original demand. Goliath and Hercules flights

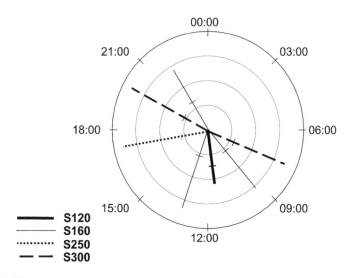

Figure 3.28 Air Hercules Before Deregulation

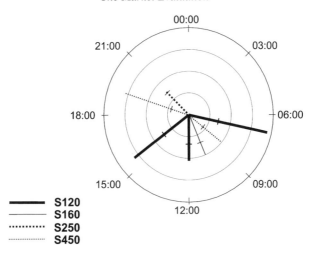

Figure 3.29 Goliath Airways After Deregulation

alternated during the day, so the majority of the spill was being exchanged between the two airlines. In general, the market was balanced and stable.

The most significant effect of the appearance of Barracuda Airlines after deregulation is that the market share and profitability of the two incumbent airlines has tumbled. Market share only improves at the expense of somebody else. By focusing on the demand peaks Barracuda would drive a wedge between its competitors, achieving an initial market share of 21.3% and forcing the other airlines' load factors down to around 60%.

The model predicts a reduction in overall spill. Although Barracuda's spill might be high, due to high load factors, Goliath and Hercules would both see less spill

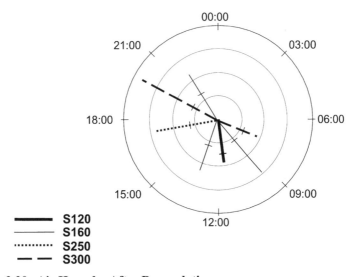

Figure 3.30 Air Hercules After Deregulation

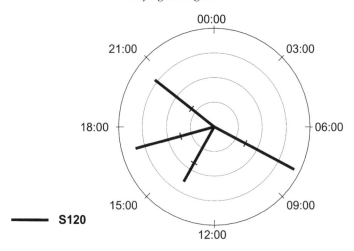

Figure 3.31   Barracuda Airlines

because of their correspondingly smaller load factors. Essentially, with more flights in the market there is an overall better matching of capacity and demand.

Another important feature of the analysis is that total passenger revenues have fallen by around 4%. This is because Barracuda have entered the market with a lower fare in order to stimulate demand.

Another way of examining the simulation results is by reference to Wagon Wheel charts, as in Figures 3.27 to 3.31.

The next phase of the analysis would be to test the various reactions of the three airlines to the initial situation. The model allows a number of scenarios to be simulated. It is a relatively easy process to modify the number of flights offered and their frequencies to try to improve the results. The initial situation presumed that Goliath Airways and Air Hercules were operating four and three aircraft types respectively. Perhaps large airlines serving a wide variety of routes with different needs can afford a fleet composition of different sizes. Our newcomer, Barracuda, might be far more limited in its ability to expand into different aircraft types, and is presumed to stick with the single-type S160 aircraft.

The model described above, termed a 'demand affectation' model, is a simple behavioural tool, that can be set up rapidly, requires relatively little input, and can be easily calibrated against observed loads. The results simulate capacity and frequency mixes for the host airline as well as the competition. As it is interactive, a variety of scenarios can be quickly tested.

The beauty of a single-sector model is that, once it has been calibrated, future demand levels can be simulated very easily. Substituting higher demand to represent future years will immediately indicate which flights need an augmentation of capacity, or where additional frequencies should be added. Once fixed, the demand curve and reallocation process should not be tampered with, unless overall market conditions alter.

When focusing on a single sector, it must always be remembered that the underlying demand may be triggered by flights that feed and emanate from the cities at either end of the route. If they change, then the demand curve will clearly change.

Modelling on a Network

Having examined a methodology for judging the appropriate capacity and frequency for a single-sector we will now move to a full network analysis. Modelling a network will magnify complexity, especially as we may well be dealing with demand on a strict origin and destination, or city-pair, basis rather than on a flight-leg basis. Firstly, there is the problem of getting equivalent and accurate data for all airlines operating in the market, which can only realistically be obtained through Market Information Data Tapes, or MIDT. Access to these data is expensive and relatively few airlines with considerable resources to purchase and then process the data will use MIDT tapes in their fleet planning. In any case MIDT only captures booking data and not final loads.

Frequently, the best quality data we can get refers only to the host airline, as the quality of data for the competition will probably be inferior to that obtained in-house. Also, it is often easier to obtain data on a flight-leg basis, rather than pure origin and destination, which is required to undertake an allocation study. Although city-pair data are ideal, only relatively large airlines possess the resources to process these data themselves. The majority of airlines solve both their revenue management and aircraft assignment problems on a flight-leg basis. This is certain to change as more sophisticated models become available. Apart from simply getting hold of the data we have another challenge in that even very simple network systems encompass a multitude of city-pair connections. These can quickly mushroom to a vast array of links. Figure 3.32 illustrates the potential city-pair connections available in our own study network.

Our fairly simple network, that may require as few as five aircraft to provide enough links for a bank of connecting passengers, offers 15 city-pair connections,

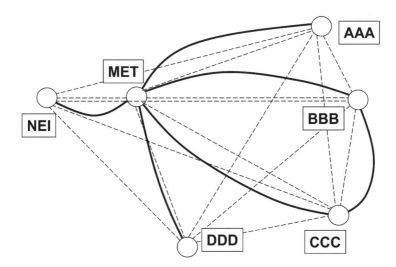

Figure 3.32 City-Pair Dependence

thanks principally to the hub at Metropolis. Each of these potential passenger journey itineraries can be made up of a variety of market segments, each with its own demand distribution by day, week or season, each with its own set of needs and expectations in terms of flight timing, frequencies, flexibility, price sensitivity, and so on. The connecting networks of complementary alliances at hub airports can potentially offer thousands of city-pair connections. No wonder their network maps look like an explosion in a spaghetti factory.

When it comes down to it, if one were to examine the characteristics of each potential passenger, the permutations would seem endless. Clearly, it would be a nightmare to try to model to this level of detail so, to avoid grey hair, we shall group the characteristics of the market into a number of categories that we shall model by using preference factors.

Preference Factors

The defunct US Civil Aeronautics Board (CAB) applied a Quality of Service Index (QSI) in market share modelling in the 1960s. The QSI encompassed a series of coefficients that explained travel behaviour. It is very useful to compress market behaviour down to a small number of independent preference factors. The trick is to find the right number. Too many and calibration of the results of the model with historic data becomes impossible. Too few and we are not reflecting enough behavioural tendencies.

Routeing preference   Firstly, passengers are known to exhibit a preference for direct routeings. This sounds fairly obvious, but one problem we are faced with is that the development of hub-and-spoke systems has forced city-pair demand into one, and sometimes, two hubs. Passengers will accept being channelled in this way because the hub-and-spoke concept creates the connections that would not otherwise be present. For example, the point-to-point demand from Neighbourhood to DDD in our model would be insufficient to justify a direct service. Yet the density of traffic from both of these points to the hub at Metropolis results in a multitude of choices.

Aircraft type preference   Secondly, passengers are known to exhibit a preference for particular aircraft types. We saw from the IATA Corporate Travel Survey that a relatively high number of high-yielding First-class travellers do actually consider the aircraft type in their airline choice. Sometimes this is manifested very simply, such as favouring a wide-bodied as opposed to a single-aisle aircraft. Sometimes the preference is more subtle, where some regular travellers have an aversion to flying twin-engine aircraft in remote areas, for example. Frequent flyers tend to get to know aircraft types fairly well, and consequently prefer some to others. However, this would only apply where the passenger really does have a choice. Schedules feature very highly in most business traveller priority lists and aircraft choice will then be less important. Also, travellers on short-haul journeys tend to be less concerned about the type of aircraft flown. One clear area where aircraft type preference plays a role is between jets and turboprops.

Travel time preference   A third area of preference is where passengers seek the shortest overall travel time. This category tends to be more important in short-haul markets where choices still exist between turbo-props and jets, and where hubbing tends to elongate the overall journey times. In fact, the concept of directing traffic through hubs, although commendable in providing high connectivity, has lengthened overall city-pair travel times. This has become especially problematic in some long-haul markets where double-hubbing exists.

Flight-leg preference   Another area of preference is where seats on a multi-stop flight might be given priority assignment to passengers travelling on all segments of the flight, rather than on local 'hops'. There are two schools of thought concerning this issue. On the one hand, a passenger occupying a seat for the full duration of a multi-stop flight bears less cost in terms of handling and ticketing. Accepting local passengers on each segment of the flight means that the risk of keeping the seat sold is higher. On the other hand, passengers flying longer distances contribute less yield, which is the measure of revenue per unit of distance flown. Long-haul passengers do contribute a higher amount of overall revenue, however. Revenue management programs are designed to resolve these potential conflicts. For the purposes of fleet planning, we do need to decide whether we would want to give non-local passengers a priority in access to seats or not.

An example of how a preference factor in demand allocation functions is shown in Figure 3.33.

On our network there are direct flights between Metropolis and CCC as well as flights via BBB. All other things being equal, that is, when passengers are faced with a choice of equivalent frequencies, the number of passengers gravitating to the indirect flight is in proportion to the two factors assigned to the two travel opportunities. Thus, if we fix the factor for the direct flight at 1.0 and that for the indirect flight at 0.4, then the total demand for the day between Metropolis and CCC, 308, would allocate according to the ratio of the factors.

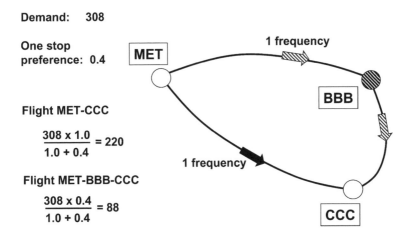

Demand: 308

One stop preference: 0.4   **MET**

1 frequency

BBB

Flight MET-CCC

$$\frac{308 \times 1.0}{1.0 + 0.4} = 220$$

1 frequency

Flight MET-BBB-CCC

$$\frac{308 \times 0.4}{1.0 + 0.4} = 88$$

CCC

Figure 3.33 Preference Factor for Direct Flights

Where life becomes complicated is that all things are certainly *not* equal. We must layer onto the problem the fact that the travel times are not identical, and the number of frequencies offered between Metropolis and CCC via BBB is only four, compared to 27 direct services from Metropolis. In a model, we must aggregate all of the preference factors together so that the effect of the direct flight preference is considered in relation to all other factors. In addition, these flights will carry other sets of city-pair demand, each with their own unique characteristics. Then, all of this demand must be compared to the capacity offered and spill calculations made.

This kind of demand allocation modelling can only be accomplished with the aid of a specialist tool. Unlike macro modelling, it is something that cannot be undertaken with a spreadsheet. The aircraft manufacturers and a number of independent software suppliers to the industry have developed such tools of varying complexity to deal with the demand allocation process on a network.

There is one further element to demand allocation that is unique to fleet planning. We must not confuse the relationship between the airline's schedule to the demand and between the schedule and the operation of the fleet. The two relationships are of necessity linked, but the strength of the link is largely a function of the time frame of our plan. Let's consider this problem.

Interaction Between Schedule and Fleet

There is a distinct difference between the allocation of demand onto a network and the optimisation of the *use* of the fleet. Demand allocation is the process of simulating how potential demand gravitates to a set of aircraft capacities and frequencies represented by the schedule, and fleet optimisation is concerned with the best assignment of a given fleet to that schedule. Indeed, the schedule is the keystone of an airline's success.

In fleet planning we are principally concerned with the demand allocation process, which is a strategic issue, than with the fleet assignment process, which is tactical. Understanding both is, nevertheless, important as they both interact with the schedule. Yet modern modelling techniques tend to focus on either one or the other. This is partly owing to the complexities of solving the totality of the problem and partly because the two issues are addressed by different parts of the organisation and with different objectives in mind. Finding the right fleet to satisfy demand is one thing; constructing a set of aircraft and crew rotations is a completely different task.

If our micro fleet plan were focused on the relative short term, let's say two to three years, we ought to attempt to integrate operational issues such as scheduling and maintenance planning into the demand allocation calculation. However, there is no point in trying to schedule a fleet that we are proposing for 10 years' hence. The relevance of building a schedule diminishes with time and we do not need to delve into too much detail for the distant future.

Nevertheless, our demand allocation methodology must take account of the efficiency of a schedule, but without actually applying real rotations into the plan. This can be achieved through another factor, called the 'schedule adaptation factor'. When our fleet plan supposes that a sector is flown by a variety of aircraft capacities we would like to reflect whether we believe that the schedule would offer high

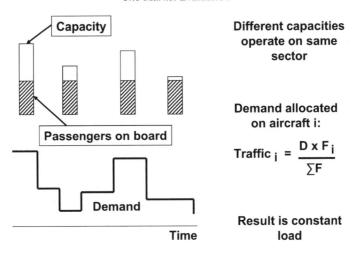

Figure 3.34 Allocation as a Function of Frequency

capacity at peak periods and lower capacities at the off-peak periods. An efficient schedule is one where there is a good match of capacity with demand.

There are two theoretical extremes that can occur, which are illustrated in Figures 3.34 to 3.37. In the examples we can see that in the case where passenger behaviour is completely indifferent to the presence of capacity, they fill up equally amongst all the frequencies offered. This results in differently sized aircraft having different load factors. Conversely, if there is a perfect match of demand to the size of each aircraft type on the route, then the load factors will be identical on each flight. In this latter case, we could conclude that the schedule offered is precisely adapted to the marketplace.

Reality will lie between these two extremes. Empirical evidence suggests that if we assign a value of zero for the frequency allocation preference and a value of one for the capacity allocation preference, we could expect to see a typical schedule adaptation factor of around 0.85. It is perfectly feasible to calibrate this factor. An

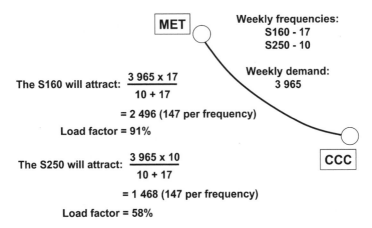

Figure 3.35 Example of Demand Allocation to Frequency

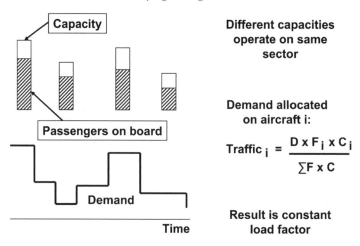

Figure 3.36 **Allocation as a Function of Capacity**

examination of recorded load factors on routes where multiple capacities are offered
will give a good indication of how well a schedule has been adapted to the market.

Having described the demand simulation process, we shall now look at how
the Goliath Airways fleet can be modelled for 2007, bearing in mind that we have
already compiled a macro fleet plan for 2007–2013 and examined their prime route,
Metropolis to Neighbourhood, using the demand affectation model.

The Micro-Network Model

Our demand simulation will be based upon a reference period of one week, with the
results being expanded to represent 2007 by multiplying by 52. We are therefore

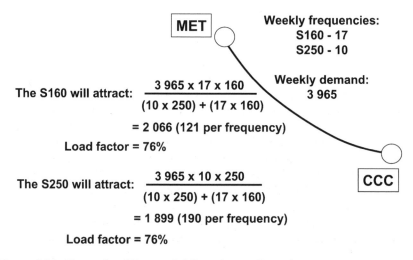

Figure 3.37 **Example of Demand Allocation to Capacity**

Table 3.7   Goliath Airways Network Simulation 2007

Aircraft type	S120	S160	S250	S450	TOTAL
Number of aircraft	4	8	3	2	17
Weekly frequencies	76	132	50	44	302
Annual utilisation per aircraft	3127	2921	2985	3181	3011
Passengers on board	384519	796120	461686	659708	2302033
Passengers on flight	384519	796120	432253	659708	2272600
RPKs (1000)	876343	1897968	1161841	1243100	5179252
ASKs (1000)	1072032	2670720	1599000	2045160	7386912
Load factor	81.7%	71.1%	72.7%	60.8%	70.1%

presuming that demand is homogeneous throughout the year and that opportunities to change capacity levels would not exist. In most fleet plans of this detail it would be prudent to undertake separate analyses for each timetable period of summer and winter, and possibly look at the 'shoulder' periods each side of the peak as well.

We have seen above that the calculations are quite sophisticated, but the input data for such a model to run are basic. The timetable, capacities of the fleet and city-pair demand are essential. Then we need to construct a set of preference factors, the schedule adaptation factor, and estimate the k-factor so that demand spill can be calculated. It is presumed that all of the demand initially spilled will be reallocated onto the Goliath Airways network. In competitive markets this is not always the case in reality. However, it is quite likely that Air Hercules, the second incumbent airline, is also generating spill that is being picked up by Goliath. This phenomenon of spill exchange has been observed in real markets. When Aer Lingus and British Airways were competing heavily on the London to Dublin route it was established that the spill generated by each airline was being exchanged and effectively cancelled out.

The results of the base year simulation, seen in Tables 3.7 and 3.8, reveal some immediate deficiencies in Goliath Airways' operation.

Firstly, the airline is probably operating too many types of aircraft. We have already seen in the macro study how the largest type, the S450, could be dispensed with in 2011. In any case, a fleet of only two aircraft is likely to be expensive unless there were to be complete commonality with other types in the fleet. The only

Table 3.8   Goliath Airways Load Factors 2007

Flight numbers Out	Back	Weekly frequencies	Aircraft	Load factor (%)	Route	
0011	0021	7	S120	79.5	MET	NEI
0012	0022	7	S450	71.1	MET	NEI
0013	0023	7	S160	77.0	MET	NEI
0014	0024	7	S120	79.5	MET	NEI
0015	0025	7	S120	79.5	MET	NEI
0016	0026	7	S450	71.1	MET	NEI
0017	0027	7	S250	73.9	MET	NEI
0701	0702	21	S160	80.9	MET	AAA
0801	0802	21	S160	55.1	MET	BBB
0901	0902	4	S250	60.4	MET	BBB CCC
						69.3 46.6
0903	0904	7	S450	48.1	MET	BBB
0905	0906	1	S250	79.5	MET	CCC
0905	0906	17	S120	83.1	MET	CCC
0601	0602	17	S160	81.7	MET	DDD
0601	0602	1	S450	78.0	MET	DDD

justification for operating such a small fleet would be if no other capacity solution could be found to deal with the relatively small number of demand peaks on the network. Our micro demand affectation study of the Metropolis to Neighbourhood route revealed that significant demand peaks did, in fact, occur in the morning and evening, and this would be enough to justify the S450 on that route.

However, in the context of the big picture, the aircraft does not seem to fit very well. On the Metropolis to DDD route there is only one frequency/week, compared with 17 offered by the smaller S160 aircraft. Critically, the S450 is not attracting sufficient loads on its daily flight from Metropolis to BBB. A 48% average load may not be generating any spill, but it is probably not generating any profits either.

A second problem apparent from the simulation is that there is a problem route. Goliath Airways are presumed not to have any traffic rights from BBB to CCC, so all of their traffic on this sector originates from the hub in Metropolis. Consequently the average load, at 46%, is rather low. Some action is now required to improve this situation.

The third issue to consider is the imminent entry into the market of Barracuda Airlines on the main trunk route to Neighbourhood. We did not take account of this in the macro plan, but as we now begin to draw the threads of all of our models together, we should consider the implications of the loss of some of Goliath's prime market.

Now that we have completed a trilogy of simulations for the base year of the fleet plan, 2007, we should ensure that the results are compatible. The detailed micro study of the trunk route should agree precisely with the micro-network study. Likewise, the network study must crosscheck with the macro study.

It would be usual to select a base year for which reported data already exists. This is because calibration is essential in order to ensure the integrity of the models and the parameters that drive them.

Applying the models to future years is the next stage. Input parameters such as preference factors should be frozen unless there is a clear reason to change them. This could be when new market segments are addressed, for example, that might be expected to behave differently to those already served. One of the main challenges in using models is that the results are highly dependent upon our predictions of market and economic conditions.

The Network Matrix

In order to plan our micro evaluation of the network we need to visualise the problem in several dimensions. It is helpful to define a network matrix, which links each unit of operation, such as the flight numbers, with each unit of time, such as a season. Figure 3.38 shows how this network matrix can be built.

In the vertical axis we can conduct a simulation of how demand allocates to the individual flight numbers, as outlined above. However, as long as our intention is to find the optimum mixture of frequencies and capacities for each of the flight numbers, we are not taking into account how needs may vary from one period of analysis to the next. If the fleet composition were to be determined for the first

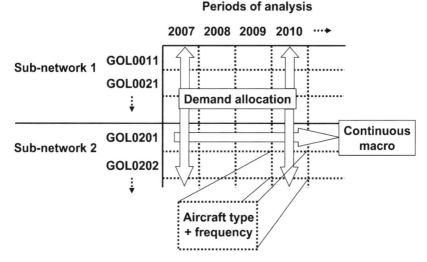

Figure 3.38 The Network Matrix

period, then this would form the starting point for the development of the ensuing period, and so on.

Building up the fleet in this way is rather like navigating a sea voyage by dead-reckoning. This is the calculation of one's position on the basis of the distance covered since the last precisely observed position, incorporating estimated corrections for changes in wind, current and compass errors. This method of navigating allows small navigational errors to accumulate as the voyage continues. In our fleet plan, we might introduce numerous small errors for one part of the network that cannot be crosschecked against the big picture.

There are other constraints at work. As time marches on, flight numbers will load until certain trigger points are reached. For example, when a planning load factor is breached, another frequency should be added or else aircraft type changed. Thus, there is a risk that the composition of the fleet could be driven by the requirements of each unit of the operation.

To get round these problems, we also need to visualise the development of the fleet in the horizontal axis. To simplify the problem, it is preferable to move from one time period to the next by dividing the network into a number of modules, or sub-networks. These sub-networks should have similar characteristics, so it is wise to separate long-haul intercontinental flights from domestic routes. In our example, we have already chosen to separate the Metropolis to Neighbourhood route from the regional operation.

Each new time period should also embrace an appropriate growth in demand. The demand forecasts should ideally be prepared on an origin and destination basis. To move forward to the next navigational position we can also expand the capacity in the fleet, either through larger aircraft, or else by increasing frequencies. The latter may also imply growing the number of aircraft of a particular size. We should therefore track very carefully the build-up of aircraft utilisation so that target limits are not breached.

Table 3.9   Goliath Airways Network Simulation 2010

Aircraft type	S120	S160	S250	S450	TOTAL
Number of aircraft	4	9	3	2	18
Weekly frequencies	70	176	56	46	348
Annual utilisation per aircraft	3210	3334	3368	3291	3308
Passengers on board	363119	1085991	539270	744525	2732904
Passengers on flight	363119	1085991	510666	744525	2704300
RPKs (1000)	915651	2501743	1357469	1399473	6174336
ASKs (1000)	1100736	3429504	1804400	2115360	8450000
Load factor	83.2%	72.9%	75.2%	66.2%	73.1%

It would be quite acceptable to move forward to the next time period by aggregating demand and supply to RPK and ASK levels. Also, if we are already looking at economics at this stage, it is easier to concoct a simplified yield function that could be applied to each sub-network rather than examine the revenue of each and every city-pair on the network. Several yield functions would suffice where different market segments are being considered.

Once we have advanced to a new time period, then another more detailed demand allocation procedure can then take place. By working in the two axes we are adopting a 'snapshot and movie' approach. Each snapshot looks at the entire operation at a single glance, and the movie makes sure that the fleet is navigated on a coherent course. This is the most efficient method of building a network analysis over time. There is no real point in analysing each and every time period over 10 years. If the fleet plan were being built by timetable period, this would mean 20 simulations. Apart from the time taken just to construct a base case, there is little point in going into what could be regarded as microscopic detail in 10 years' time. It is much better to analyse perhaps the first two to three years in detail, and then jump forward by applying an aggregated technique to a point in the future, and so on. This technique, called 'mile-posting', is a reasonable compromise between complexity and analysis time.

Our analysis now comprises two snapshots, in 2003 and 2006, the results of which are seen in Tables 3.9 and 3.10.

The traffic growth and fleet sizes are consistent with the macro plan and the results of the demand allocation are almost exactly in line with the limiting dimensions we set. So, load factors have stabilised at around 75% and the utilisation of each aircraft is approximately 3,300 hours. When making a final tally of load factors, utilisation, ASKs and RPKs, it is essential to ensure that the progression of units and seats is a smooth one and in line with forecast growth.

Table 3.10   Goliath Airways Network Simulation 2013

Aircraft type	S120	S160	S250	TOTAL
Number of aircraft	8	9	6	23
Weekly frequencies	136	182	130	448
Annual utilisation per aircraft	3156	3367	3315	3280
Passengers on board	640840	1124942	1269152	3034934
Passengers on flight	640840	1101195	1269152	3011186
RPKs (1000)	1624448	2574485	2670573	6869506
ASKs (1000)	2164032	3463616	3551600	9179248
Load factor	75.1%	74.3%	75.2%	74.8%

The process will not end here. We have not considered the impact that Barracuda Airlines might have on Goliath's results, should they survive. Neither have we restructured the network to deal with the potentially problematic route of BBB-CCC. Indeed, a variety of sensitivity tests are called for in order to determine the best fleet mix.

When building up the fleet over time, the analyst must consider factors that are beyond the optimal matching of supply and demand. Practical issues such as the number of aircraft types in the fleet may dictate the fleet build-up. The most efficient fleet is the one that is streamlined in terms of numbers of types and draws maximum benefit from families of design. So, in our example, we withdrew the S450 from the fleet. Also, the inclusion or exclusion of a particular aircraft from the fleet may be driven by financial considerations, such as the availability of capital to invest, or the opportunity of realising a significant book profit by disposing of an aircraft and replacing it with another. Such extraneous issues emphasise once again the fact that fleet planning should never be driven by computer programs alone.

In this Chapter we are discussing the main fleet planning modelling approaches, but by no means the only ones. The models shown have been designed to be applied to many different types of scheduled airline. Non-scheduled carriers, cargo operators, low-cost carriers and leasing companies will all have very different planning criteria. For example, leisure airlines must focus on behavioural patterns driven by the tourist industry. Cargo demand is driven by trade and the degree of directional imbalance in the market. Low-cost carriers will deploy their capacity as close as possible to the optimum operating range and set pricing levels so that the market gravitates to their schedule. Leasing companies place far less emphasis on how a fleet performs in an airline but consider the aircraft as an investment with a residual value potential.

Whatever the business of the airline, analysis of markets is essential. Although the composition and growth of those markets might differ dramatically, according to geography, the economy and many other factors, understanding their dynamics is fundamental. Yet market analysis is very much a grey area, with no right or wrong solution. The best we can hope to achieve is an appreciation of which elements shape the demand and how to influence them. At this stage of the fleet plan we have now ironed out roughly what kind of aircraft might be appropriate for the market we wish to be in. However, we have yet to consider the particular aircraft types and the economic implications of the fleet mix that the market might require.

Before considering the aircraft evaluation itself, there is one final piece of the market jigsaw to consider: market share.

Market Share Modelling

Our micro model of the Metropolis to Neighbourhood route enabled us to determine, for a given demand curve and spill model, how demand might allocate to individual flights. We created, by default, a market share model. In practice, an airline's market share will be influenced by a vast array of factors. These will include the number of frequencies, ticket price, the extent to which capacity is available to coincide with the underlying demand, aircraft type, the perception of the airline's overall product,

Assume:
Airline 1 offers 6 frequencies
Airline 2 offers 10 frequencies
Exponent to be tested: 1.8

Frequency share of airline 1 = 37.5%

$$\text{Market share of airline 1} = \frac{(\text{Frequency ratio})^{1.8}}{1 + (\text{Frequency ratio})^{1.8}}$$

$$= \frac{(0.6)^{1.8}}{1 + (0.6)^{1.8}} = 28.5\%$$

Figure 3.39   S-Curve Example Calculation

and so on. In fact, modelling all of the elements simultaneously is an impossible task, if only because many elements may be correlated with each other. Also, the determinants of choice have a different ranking depending upon whether we are looking at short or long haul. Short-haul business travellers tend to be more sensitive to a difference in scheduled departure or arrival times than long-haul business travellers.

As you might imagine from the complexities of market behaviour it is very challenging to build an all-singing, all-dancing market share model. There are, however, more straightforward models that can address specific elements of the problem. We shall now examine two approaches.

The S-Curve

In the Metropolis-Neighbourhood example we presumed that demand was sensitive to frequencies. This may be the case for corporate business travellers, although the price-sensitive independent business and leisure markets are less likely to be primarily influenced by frequencies. Nevertheless, there is a mechanism that we can use to establish the relationship between the amount of frequencies offered and market share, based on an S-curve relationship between the two factors. Put simply, where two or more airlines compete, the airline that offers less than half of the frequencies can be expected to gather a proportionately smaller portion of the market. Conversely, as frequency share predominates then that airline can expect to gain a proportionately greater share of the market.

An S-curve can be established for a market by examination of how on-board loads evolve as a function of a change in the frequencies offered. A typical curve is shown in Figure 3.39, where the shape is determined by the ratio of frequencies offered and an exponent, in this case 1.8. The exponent can be calibrated according to observed market data.

One problem that we often encounter in fleet planning is in getting different models of the same market to agree with each other. Incompatibility of results is

Market share (%)

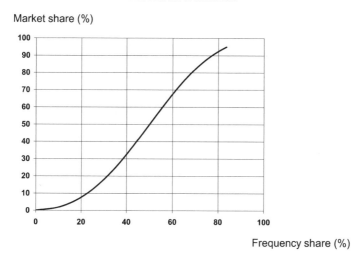

Frequency share (%)

Figure 3.40 S-Curve Relationship

not necessarily a problem because different models take into account different parameters. For example, in the Metropolis to Neighbourhood demand simulation Barracuda Airlines' four frequencies might be expected to yield rather less than their proportion of the traffic. In fact, this is not the case. In return for only 17.5% of the capacity and 22.2% of the frequencies, our simulation suggests that they might achieve 21% of the market. Not much of an S-curve here!

However, what the S-curve is not taking into consideration is the ability of Barracuda Airlines to offer flights at the peaks of the demand curve. The better matching of supply and demand has mitigated the S-curve effect. The lower price offered by Barracuda also attracts demand. One should not underestimate the typical consumer reaction of try-it-once, as well. Just because the S-curve does not appear in our model does not invalidate its principles.

The S-curve can work in certain markets, but not others. For example, the relationship between market share and frequency share can only hold true so long as the market is indeed sensitive to frequencies. Thus, leisure markets are less likely to respond to frequency.

Also, the tendency to direct demand through hubs in order to improve network connectivity means that the frequencies along the feeder routes, or spokes, to the hub are determined not by the level of demand along those spokes, but rather by the timing of the hub connections themselves. This means that the aircraft size employed at any moment of the day does not necessarily correspond to the demand at that time.

The S-curve relationship works well in short-haul high-frequency markets but has less relevance for long-haul routes where it is more difficult to build frequency as a competitive weapon.

An overall airline development objective can revolve around a critical size target. This is contingent upon the moment at which an airline enters a market. Start-ups tend to need to establish a critical mass as early as possible, and the degree to which

Assume:

Exponents to be tested: a = 1.4

b = 1.0

$$\text{Flight attraction} = \frac{(\text{No of weekly flights Airline 1})^a}{1 + (\text{Average weekly stops Airline 1})^b}$$

	Non stop	One stop	Two stops	Total flights	Average no of stops	QSI	Market share
Airline 1	2	2	1	5	0.80	5.29	47.1%
Airline 2	3	1	1	5	0.60	5.95	52.9%
Airline 1	2	3	1	6	0.83	6.71	49.7%
Airline 2	3	2	0	5	0.40	6.79	50.3%

Figure 3.41   QSI Example Calculation

this is important depends upon whether a start-up is competing directly with a large incumbent or whether a new niche market is being addressed.

Finally, the type of market being served is important. Business markets tend to be more sensitive to the provision of frequent departures than leisure markets. The S-curve is a useful, but not the only guide, as to how market share can evolve as frequencies develop.

Quality of Service Index

Another commonly applied and straightforward market share technique is based on the Quality of Service Index (QSI) originally promulgated by the former US Civil Aeronautics Board. We have already seen elements of the QSI when we worked through our micro-network plan, in the form of preference factors to determine the allocation of demand. The QSI enables us to model passenger behaviour in terms of their preference for minimum travel time, minimum number of stops on a route, frequency of service and airline image.

In Figure 3.41 we can see an example of how the calculation is made. As with the S-curve, it is essential to calibrate the calculation with historic data by means of exponents. The example shows two airlines that compete on a route, offering a mixture of non-stop, one-stop and two-stop services. When both airlines offer the same number of services, the higher proportion of non-stops might be expected to attract higher demand. What the model shows is that even if Airline 1 actually increases the number of services on a route, it is not sufficient to capture the greater market share so long as the number of non-stops offered remains inferior to those offered by Airline 2.

More sophisticated models are, of course, available in order to predict the effect on traffic volumes and market share of changes in route structure, service level, schedule and price. For example, the logit model is a popular probabilistic model representing the discrete band of choice behaviour of individuals. Logit models are useful because they react very quickly to ever-changing environments and give accurate results. They are used in flight scheduling and in revenue management systems. However, simpler QSI models remain attractive due to their transparency and ease of use. Air New Zealand, for example, performs its origin and destination demand forecasting with a QSI-based system.

In Summary

In this Chapter we have examined the importance of market analysis to the fleet planning process. The necessity of taking a long-term view in fleet planning, coupled with the problems of accurately predicting how markets will evolve in highly competitive situations, means that we need to adopt several parallel but complementary approaches. Thus, macro and micro modelling should go hand-in-hand. Each requires a different set of inputs, each produces results of different complexity, but both models should agree with each other.

To illustrate the basic structure of a fleet plan, we constructed a macro plan with a six-year horizon, then a micro plan to examine an important route in the network, and finally a second micro plan to consider the entire operation. These models can act as building blocks to interpret the dynamics of market behaviour. They are not by any means the only models available, although the approaches outlined are in consistent use throughout the industry. These building blocks can be supplemented by a series of additional models, where appropriate. Market share models, such as the S-curve and QSI, are examples of these. However closely we believe we have assessed the fleet requirement to serve our market we should always remember that models never, ever, tell the whole story.

A marketing analysis can reveal whether the best solution is to opt for high frequencies at the expense of large-sized aircraft, whether underfloor cargo space is essential, whether aircraft of particular range ability are needed, and so on. Our marketing analysis cannot advise which aircraft model should be selected, or the conditions under which the aircraft should be acquired. The next phase of our fleet plan is therefore concerned with identifying aircraft attributes and matching them to the airline product.

Chapter 4

The Aircraft and Airline Product

Getting Prepared to Conduct the Evaluation

It is vital that the fleet planner has a good appreciation of the product strategies of the major airframe and engine manufacturers. In this Chapter we shall map out the competing offers, together with the history of product development, with particular focus on commonality and family benefits, which have become industry watchwords for maximising fleet efficiency and economics.

We shall then move on to examine how changing expectations of the airline product have been a catalyst in the development of a more adaptable aircraft cabin design. We shall see that there are conflicting schools of thought. One idea is that the aircraft should be a highly-customised tool to enable delivery of an airline brand. The other idea is that the aircraft cabin is a simple commodity. We must also examine some key issues in the customisation process – at least, for the purposes of aircraft evaluation.

We shall finally review the various documentation needed in order to conduct an aircraft evaluation, with a review of documents provided by both the manufacturer and the airline and, critically for the entire process, we shall look at the compilation of the ground rules, or study assumptions.

The Rough Cut

In order to form a better impression of the airline product and how it impinges on the aircraft type, or vice versa, it is useful to draw up a shortlist of potential aircraft which could fit the bill. There are so many variations on offer that it is sometimes prudent to start with a 'rough cut' of any potential solution and run through a quick analysis to eliminate obviously unsuitable contenders. Manufacturers helpfully supply summary information on aircraft configurations, weights and ranges so that we can quickly get a feeling for whether a particular variant is likely to fit the network or not.

Sometimes it is a technical limitation that can force the early retirement of an aircraft type from the evaluation. Particular airfield constraints, such as short runways, or else *en-route* conditions, such as high terrain or an absence of diversion airfields, may immediately dictate whether attention should be devoted to either a twin or multi-engine solution.

It is at the rough cut stage that we really need to refer to our categories of aircraft selection criteria as outlined in Chapter 2 (Figure 2.4) to ensure that we set off on the right track.

The rough cut can more easily eliminate contenders in a fleet plan for an existing operation than for a start-up. A clean sheet of paper has numerous advantages, but

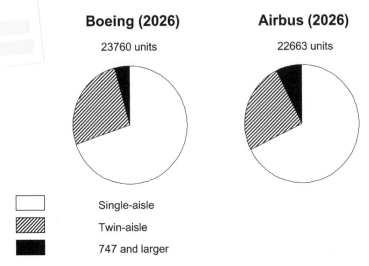

Figure 4.1   Global Unit Forecasts for Passenger Aircraft
Source: Derived from manufacturers' forecasts. Excludes regional jets

homing-in on precisely the right aircraft to analyse is more difficult, especially if the intention is to serve routes that have never been linked before.

In order to make this preliminary selection we need to review all of the aircraft products on offer, both now and in the future.

The Aircraft Product

By the time we start considering specific aircraft types for the fleet plan we should already have covered a number of analytical stages. We should know our market, our competition, our future traffic prospects, and we have defined our brand. Now we need to get to know the suppliers and their products.

Today, the large civil airframe manufacturing business is in the hands of two players, Airbus and Boeing, and the market is fairly evenly split between them. Both have tremendous strengths and a very full product range, although there are some quite fundamental differences in approach in some areas, as we shall see.

Evolution of Aircraft Demand

The 20-year forecasts of Airbus and Boeing, called the Global Market Forecast and Current Market Outlook respectively, agree that the overall growth rate of RPKs will continue at just under 5% per year over the next 20 years. The two manufacturers differ in how they believe this demand will be served. In essence, Boeing believes that growth will gravitate to smaller aircraft, whereas Airbus envisages a stronger market for large aircraft, specifically the A380.

Global trends do not, of course, give any indication of what the needs of an individual airline might be. However, manufacturers' perspectives do give a clue as to how the general aircraft market is likely to develop in the longer term. It is useful to know whether we might be committing ourselves to a size category of aircraft that might be declining or increasing in magnitude. Boeing and Airbus differ on this rather fundamental point. Boeing believes that international growth will be served by more point-to-point flights, relieving congested hubs. Airbus, on the other hand, contends that many dense routes will not only result in fragmentation, but will need very large aircraft in addition.

Adherence to these quite different perspectives inevitably leads to differing views on aircraft size needs.

Product Line Development

Both Airbus and Boeing now offer product lines covering virtually any configuration, and it is becoming easier to second-guess where attention will be focused in the future in order to maintain homogeneity of technology.

Historically, Boeing had a head start over Airbus, having already developed a significant number of aircraft types before their European rival became established in the 1970s. At that time Boeing's stable included the early long-haul aircraft of 707 and 747 types, along with short-haul 727 and 737 types. The Lockheed 1011 TriStar and McDonnell-Douglas long-haul DC-8 and DC-10, and short-haul DC-9 aircraft completed the main picture, along with smaller offerings from Fokker. When Airbus entered the market with their twin-aisle twin-engine concept, not many in the industry gave them much chance of success. Perseverance and the development of complementary families of aircraft paid off for Airbus, whereas McDonnell-Douglas' failure to develop a coherent product range ultimately contributed to their demise.

Two broad philosophies emerged from the remaining two players. Boeing built upon their proven technologies by developing a series of highly successful derivatives, with the 737 Next Generation family and 747-400 developments being good examples. Boeing also launched themselves into new categories of aircraft in the 1980s and 1990s. The 757 and 767 plugged the gap between the 737 and 747 and the various 777 mid-market models have been an undoubted success for Boeing. Boeing also incorporated the former single-aisle MD-95 of McDonnell-Douglas into their product line, cleverly rechristened as the 717.

Airbus, on the other hand, invested in new technology from the very beginning. The backbone of the early years of Airbus, the A300B4, was developed into the A310 and A300-600 types. The development of the single-aisle fly-by-wire A320 family, intended as a replacement of 737, 727 and DC-9 types, turned out to be a resounding success for the European manufacturer, with the technology concepts being fed through to the longer range and wide-bodied A330 and A340 families as well.

The next major phase of development took place in the 2000s. Airbus spent the first part of the decade with considerable focus on the high-capacity long-range market, served by the A380. In a sense, this aircraft marked the completion of the Airbus product range, and numerous variants were now on offer within each main type, offering a broad spectrum of seating capacities and performance capabilities.

Similarly in the early 2000s, Boeing concentrated on further developments of both the 737NG and the 777 into families of aircraft. So, both Boeing and Airbus were able to address all segments of the market at this stage. Nevertheless, things were about to change.

The first major departure from traditional thinking came from Boeing who offered what turned out to be a still-born product, the Sonic Cruiser. The concept was a long-range 300-seater that could fly at up to Mach 0.98 and radically reduce journey times. Reductions of 75 minutes on trans-Atlantic flights and around three hours on trans-Pacific flights were quoted. It was thought that such an aircraft would be attractive to time-sensitive travellers and also improve fleet productivity. The project failed, partly because of the fall-out from the 11 September 2001 terrorist attacks, and partly because it was established that passengers may not, in fact, be willing to pay a premium for speed. Environmental issues, such as fuel burn and noise, and certification issues as well, contributed to the uncertainties. What the Sonic Cruiser did was shake up the airframe manufacturing business and jump-start some more innovative thinking.

The next major innovation from Boeing was the 787, destined to become a powerful competitor in the 'middle market' in terms of large aircraft seating capacity. Boeing needed to find a cost-effective replacement for the 767 and, ultimately, the 777. The concept of the Dreamliner captured the imagination of the public and airlines alike and the aircraft looks certain to have a bright future, especially as it is offered as a true family – something that Airbus has been promoting as the way to go for decades.

It was very clear that the 787 provided Boeing with a totally new technology platform, which it may ultimately use to replace the 737NG, when the timing is right.

In parallel to these dramatic awakenings in Boeing, Airbus was facing a similar challenge, in that it needed a radical rethink to update its own mid-size aircraft models. The A300-600 market, like that of the 767, had reached the end of its natural product life. The smaller of the A330 variants, the -200, was selling very well, but it was clear that Airbus' own product range would need to be refreshed in the 2010s. The chosen solution was the A350.

Airbus suffered from a misperception that the A350 was a warmed-over A330 at first. Indeed, it went through four iterations of design before the A350 began to attract sufficient interest in the market. To answer the needs of the market Airbus relaunched the design as the A350XWB (extra wide-body), providing the airlines with a real alternative to the 787.

In addition to the main Boeing and Airbus products described above, there are numerous offerings for the cargo market. The MD-11, successor to the somewhat troubled DC-10, has outlived its passenger days and has becoming a predominantly freighter aircraft, many having been converted from a passenger role. Similarly, an increasing number of early A300B4 aircraft are also finding a new role as cargo aircraft. New pure freighters are also in the product lines of Boeing and Airbus.

The manufacturers have also turned their mid-market aircraft of both 777 and A330 types into pure freighters. Boeing has prolonged the life of the 747 yet again by coming up with the stretched 747-8, available in both a passenger and freighter variant.

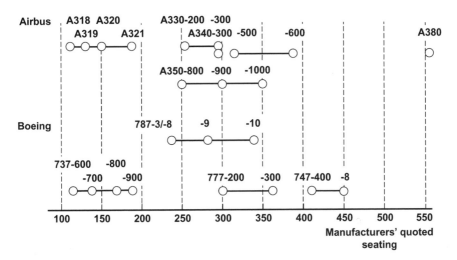

Figure 4.2   Competing Product Lines
Source: Manufacturers' generic data
Single aisle: 2-class, Twin aisle: 3-class

Both Airbus and Boeing have addressed the corporate business market by offering VIP versions of their most popular products. The Airbus A319CJ (Corporate Jet) vies with the various 737-derived BBJ (Boeing Business Jet) for this small but lucrative niche market. In addition, both manufacturers can offer most of their products in a VIP configuration.

The manufacturers will always continue to improve their product lines, both in the spirit of competitive innovation and to maintain healthy sales and prolong the life cycle. Plans for the A320 Enhanced include a new cabin interior, weight reduction initiatives, avionics improvements to enable more accurate navigation, and head-up displays to enable better airfield access during bad weather. This package is expected to deliver at least a 4% performance improvement and stave off the need for replacement. In any case, no engine is yet available to power a totally new design.

This does not mean that the engine manufacturers are idle. CFM is developing new technology for its CFM56, delivering lower fuel burn and maintenance costs and IAE is introducing the V2500 Select, with similar offerings.

The Family Concept

The development of families of aircraft sharing the same design philosophy, systems, flight decks and handling characteristics came very much to the fore in the 1990s. On the one hand, keeping the same basic design and producing variations with different fuselage length and take-off weights has great virtue in minimising manufacturers' production and development costs whilst expanding sales opportunities. Also, commonality between aircraft of different sizes and applications offers significant synergies to the operators of their products. Not only are enormous cost savings available but risk levels are also better understood and reduced.

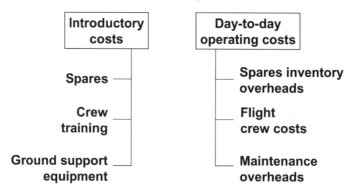

Figure 4.3   Fleet Commonality Savings

There are different forms and degrees of commonality which contribute to quantifiable cost savings. These can be conveniently divided into introductory costs and day-to-day operating costs. It is important to segregate cost savings which become available due to pure economies of scale from those associated specifically with commonality in design and operation. Figure 4.3 summarises the latter.

Cost Savings Through Commonality

Spares provisioning   On paper there are significant potential savings, with almost total theoretical airframe parts commonality for aircraft within a family of the same year of build.

In practice, it is difficult to realise the full potential because many parts have sporadic demand and modification standards vary to a significant degree. Thus, an aircraft delivered today may have less in common with a three-year earlier delivery standard than with a contemporary member of the same family. This means that commonality is very much time dependent. The real value of the family is through familiarity with the technical publications and maintenance system, and vendor relations. Even the scale effect of initial provisioning is relatively small compared with the potential effects of improved repair cycle times or improved delivery lead-times from vendors.

Ground support equipment   All types in the fleet obviously originate from the airline main base, but outstation savings are a function of intensity of use and the degree to which aircraft from the same family serve the same routes. Another factor that comes into play is whether or not other airlines have already invested in compatible ground support equipment that may be available for use.

Crew training   Training is a huge area of potential saving, especially for flight crew. Airlines that operate multiple types have to manage separate groups of pilots, each of which is subject to recurrent training. When a pilot migrates to another aircraft type with no commonality he or she must undergo a transition training which consists of up to a full 25-day type-rating course. Nevertheless, the duration of the transition can be significantly reduced through intelligent aircraft design. Thus, the 777 and 787 were designed 15 years apart and it is clear that technology has evolved.

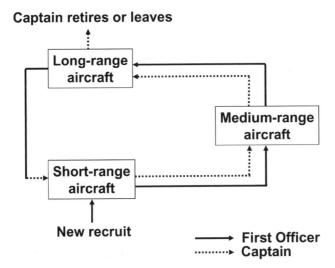

Figure 4.4   Traditional Crew Progression

The design challenge is to capitalise on technology whilst retaining a degree of familiarity between the flight decks. Thus, the 787 has a digitally recreated feel and functionality of the 777, albeit with more advanced systems that are invisible to the crew. In this way, the planned transition time from non fly-by-wire Boeing flight decks to the 787 could be as little as 10 days.

Progression through a hierarchy can result in a very heavy training programme, even when only three distinct types are operated, as seen in Figure 4.4. Thus, an airline operating three independent aircraft types would need to process six transitions whenever a Captain retires or leaves.

Commonality of flight deck design and function gives various advantages. A Same Type-Rating (STR) is available for aircraft having the same Type Certificate. A new or derivative aircraft that has functional equivalence and similar handling qualities to another family member can be awarded an STR. In this case training is valid for any member of the particular class, with only a one or two-hour briefing necessary to make a transition.

Cross Crew Qualification (CCQ) is an Airbus designation for the process of qualifying for a new aircraft type of another family by focusing on the differences between the new and currently-qualified aircraft, rather than undergoing a full type-rating for the new type. The normal 25-day type rating can be reduced to as little as a single day, depending upon the transition.

Mixed Fleet Flying (MFF) is the operational practice where one pool of pilots is permitted to fly different aircraft types concurrently. There are numerous benefits associated with this practice. Pilots are able to combine short-haul and long-haul flying, which means that pilots habitually involved in long-haul operations are able to handle more landings owing to the more intensive nature of short-haul operations. This practice can ease 'currency' training, which is necessary if a prerequisite number of take-offs and landings have not been performed within a 90-day period. For example, this might mean accomplishing three take-offs and three landings on

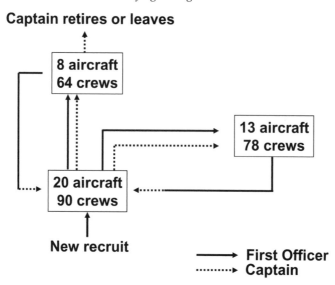

Figure 4.5   Number of Crews with No Commonality

either aircraft within this period, with at least one take-off and landing performed on each aircraft. Landings in an approved simulator are acceptable.

To get an idea of the value of flight crew training cost savings through commonality, let's take a base practical case of 20 short-range, 13 medium-range and eight long-range aircraft, and a pilot attrition rate of between 2% and 3%/year. This would equate to an annual type-rating requirement of 31 crews. Along with recurrent training for 232 crews, the total flight crew training cost could amount to around $7.7 million per year. Figure 4.5 shows the structure of this fleet and the number of crews required per aircraft.

By introducing CCQ between the long-range and short-range aircraft, plus MFF so that the long-haul crews can fly the short-haul aircraft, then 15 fewer crews would need to be trained each year. Not only are training costs reduced, but there are also savings because fewer crews are needed and overall crew productivity is improved. The productivity gains come from two directions. Firstly, the proportion of flying duty increases because less recurrent training is required and, secondly, the amount of time spent on reserve duty falls, because reserve crews are capable of handling more aircraft types. Also, where fleet assignments are such that type-specific crews might normally be 'stranded' somewhere in the network, MFF ensures that the crews keep moving because they can be assigned on another type for which they hold qualification. The savings in this scenario would amount to $1.5 million per year for training, and a further $3.2 million for payroll. Figure 4.6 summarises the concept.

An increasing number of airlines are now applying the twin concepts of CCQ and MFF, and are realising annual flight crew savings per aircraft of between $250,000 and over $1 million.

Hand-in-hand with aircraft and flight crew commonality comes 'dynamic fleet management', which is a flexible approach to aircraft assignments according to changes in demand levels. This concept shall be explored in more detail in Chapter 6.

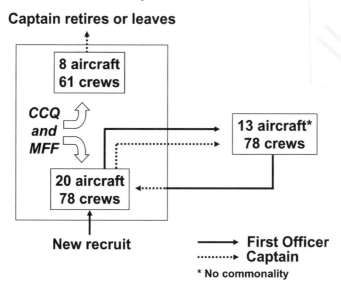

Figure 4.6   Number of Crews with Commonality

Engineering support costs   Commonality advantages extend beyond the flight deck. Engine families of a similar technology generation share a high degree of commonality, too, although airlines have placed less emphasis on this aspect of the aircraft. Training for engineers is likewise reduced as a consequence of commonality of systems and procedures. Spares inventory can be reduced, as many components within an aircraft family are common. Maintenance overheads are also lower as a result.

Also, flight deck commonality lends itself to reduced investment in simulators. It is possible for a single simulator to emulate a series of ostensibly different aircraft that share an identical flight deck layout. When fleet economics are being assessed at the investment appraisal stage, advantages accruing from lower simulator investment should be reflected.

Combining all of the introductory and day-to-day commonality savings influences fleet planning to a significant degree. It has to be admitted that an airline should not compromise heavily of the size of the aircraft it really needs to serve its market but where a choice of model exists, there is no question that lifetime costs are reduced thanks to the family concept.

The Airline Product

There is a school of thought that suggests that the airline industry is becoming rather soulless, with air travel in danger of becoming just another commodity. It is certainly the case that the low-cost carriers have built themselves a substantial niche in the market by promulgating that idea. What keeps the commodity notion at bay is branding and the constant need for innovation and differentiation. Airlines love to reinvent themselves, and when they do they spend up to, and sometimes beyond, ie price of an aircraft in designing and implementing a new image. British Airways invested €150 million in 2006 to upgrade Club World, no doubt encouraged by a 12% upsurge in premium traffic the previous November. United Airlines planned to spend $165 million on a Business-class seat improvement. This investment was in part motivated by American Airlines and Delta Air Lines' own upgrades. Malaysian Airlines invested RM700 million to upgrade 32 747-400 and 777-200 aircraft with new First and Business-class cabins and a new in-flight entertainment (IFE) system throughout. Leading aircraft cabin interior manufacturer B/E Aerospace reported revenues of $1 billion in 2006.

It is very clear that this is an issue of huge importance to the fleet planner. We need to understand how the design of an aircraft allows the airline to deliver its brand, and how the projection of the brand affects the performance of the aircraft. Performance will be affected due to weight and configuration changes, which impact unit economics. To make these judgements we should identify those elements of a brand that relate to the air travel portion of the overall airline experience. Let's review the evolution of the aircraft cabin and consider major recent developments.

How the Cabin Product has Evolved

When 747s began to replace 707s in the early 1970s airlines were suddenly confronted with so much cabin space they hardly knew what to do with it. Lounges in the sky were the order of the day, even for lucky Economy-class passengers on American Airlines, which also included the facility in their DC-10s. When economic realities began to bite, and traffic growth accelerated, airlines quickly crammed more and more revenue-generating seats into the fuselage. The A380 provides significantly more space for airlines to play with and although certain airlines certainly offer more amenities for their passengers, most operators prefer to capitalise on the opportunity of lowering their unit cost by pushing up the seat count.

Premium passengers continued to enjoy enhanced comfort levels in First and Business-class cabins throughout the 1980s and 1990s until product improvements reached the point where the quality of the Business cabin even exceeded that of the previous-generation First-class.

This trend has been accompanied by several problems. There is an increasing risk of First-class passengers migrating to the Business-class, an issue that can be contained by distinct brand differentiation. Also, with improving comfort standards for all premium passengers there is less and less space left for the higher volume, albeit lower yielding, Economy-class market.

As we saw with our market segmentation analysis in Chapter 3, the business market is dividing into two categories: the corporate market, a portion of which will always be willing to pay for First-class; and the independent business market, which is more receptive to reasonable quality at a price pitched between Economy and Business-class. Similarly, the Economy-class market is sub-dividing. On the one hand we still have the traditional leisure market, which is highly price-elastic and will always seek the best deal. However, we have now seen the emergence of the Premium Economy cabin, available to passengers willing to pay either the full Economy fare or else a supplement in order to get access to those extra inches of seat pitch.

Airlines have similarly diverged in their strategies to serve the premium market. We have seen one group of airlines abandon First-class altogether, such as KLM, Continental Airlines and Delta Air Lines, and another group reaffirm their commitment to this marginal but highly lucrative market, such as British Airways, Air France and Japan Airlines. Yet it is increasingly the case that a form of enhanced Business-class enables high-profile airlines to serve their premium markets more competitively.

A major development has been the rapid evolution of cocoon seats in premium classes, which provide 180 degree lie-flat positions within a fixed shell. This type of seating has become very rapidly deployed in long-haul fleets in all regions. It has had the effect of giving the Business-class passenger a standard of comfort unimaginable in the last century.

Welcome to the Arms Race

An arms race may be defined as an escalating competition between two nations in order to achieve military supremacy. Both sides deliberately attempt to out-produce the other in terms of larger armies and more firepower. As we experienced during the Cold War, the result of an arms race is a ridiculous excess of capability. The writer Carl Sagan had a nice way of putting it. He said that it was just like 'two men standing waist-deep in gasoline; one with three matches and one with five'. There are no two ways about it; airlines have engaged in an arms race in the cabin in order to stay ahead of the game.

Let's start by reviewing passenger needs and expectations, according to different market segments.

Space and comfort   Passengers have different needs. Whether one craves for privacy or else an environment of companionship, everyone will want a reasonable amount of space. Indeed, passengers travelling in the premium classes will expect it.

The sardine philosophy of 'pack 'em in and pack 'em tight' has thankfully given way to more reasonable configurations for all classes in the aircraft cabin. After all, the human body can endure only so much punishment, and important strides have been made in the gradual reduction in the torture and physical risks associated with being confined in an aircraft seat for increasingly long periods. Technical improvements in seat design have resulted in improved body support and more legroom. all new aircraft types must be equipped with seats able to withstand forces equivalent to 16g. These should remain in their original fixed position when subject to forces from any axis. The rigorous application of fire-resistant materials, such as fire-blocking seat cushions and light alloy structures, has gone hand-in-hand with more comfortable seat design.

A key parameter is the seat pitch, or distance between the back of one seat and the seat behind. Passengers understandably clamour for more seat pitch and wider seats. For the airlines this obviously means less seats in the same space. Despite these welcome moves, a number of charter operators persist in offering seat pitches as low as 28 inches, which is the legal minimum.

The medical profession oscillates from the notion that cramped seating magnifies the risk of fatal blood clots, to the idea that there is no evidence that Deep-Vein Thrombosis (DVT) is related to seat pitch. The British Medical Association issued a report in 2004, 'The Impact of Flying on Passenger Health', based on more than 100 scientific studies. Its conclusion is that air travel is no more harmful than sitting in front of a computer for a long period of time. Certainly, the report emphasised that passengers should engage in exercise and walk around the cabin to reduce the threat of DVT. Airlines do need to make efforts to counter the negative perception that DVT is linked to air travel – the so-called 'economy-class syndrome'. Pressure may mount for airlines to provide more circulating areas in the aircraft, which may have an effect on seating density and, hence, aircraft economics.

Although at first glance it might appear that reducing seat count is simply reducing revenue-generating opportunities, it might be a blessing in disguise in some cases. In any situation of over-capacity, taking seats out of the airframe would be less dramatic than changing the aircraft itself or reducing frequencies. Another advantage is that on long missions where the aircraft is payload-limited a reduction in seat count helps balance passenger against cargo loads.

A loss of seats to improve comfort can be compensated in other ways. One alternative is to capture higher loads, which may defeat the objective of giving each passenger more individual space, as well as increasing the risk of demand spill. The second alternative is to charge more for the extra comfort. Even price-sensitive leisure travellers are prepared to pay a premium for better comfort on long-haul charter flights. A third solution is to maintain the cabins exactly as they are and change the fare policy. This interesting twist on the problem is what China Southern did when they dispensed with First-class on their North Pacific routes. They simply re-designated their First-class as Business-class, with the Business-class being charged at the full Economy-class fare. The aircraft configurations did not change, but yields declined. China Southern thus became the first carrier on the North Pacific to abandon First-class.

Seats abreast   Another aspect of the airline product connected with the dimensions of the aircraft cabin concerns the number of seats which can be accommodated abreast. In the interests of economics, it is advisable to fully utilise the cross-section of the fuselage of the aircraft. In the interests of the airline brand, there are distinct limits to the number of seats abreast, especially in the premium cabins. The battle cry 'no middle seat' has been taken up by the airlines as well as manufacturers in their product promotion. The only really acceptable Business-class configuration is one that does not entail selling the middle seat of a triple arrangement. With the so-called '2-2-2' (three sets of double seats in a twin-aisle fuselage) becoming the norm, this has implications in the way in which an aircraft cross-section can be efficiently filled. A similar problem occurs in the high-density part of the cabin where a '2-5-2' arrangement has fallen out of favour, as this necessitates the person assigned a seat in the central section having to clamber over two persons on each side.

It is very obvious that the varying fuselage dimensions of long-haul aircraft adapt rather differently to the number of seats abreast and this remains a contentious and difficult area to resolve.

Boeing estimate that the seating capacity of the 787-9 would rise from 259 to 280 seats if airlines configured to nine-abreast instead of eight-abreast seating. With two-thirds of 787 customers planning to configure in this fashion it seems clear that the debate is set to rage on.

In truth, there is no right or wrong here. However, the argumentation of the manufacturers does need to be interpreted alongside a vision of what is acceptable for the markets being served by the airline.

The Economy-class cabin   Although unkindly referred to as 'sardine or cattle class' this cabin is of vital importance in view of the market size and the fact that the leisure segment is forecast to grow at a faster rate than the higher yield segments. There is a very clear distinction between the attitude of long-haul airlines and short-haul airlines concerning the Economy-class cabin. Long-haul airlines focus on providing 'just enough' comfort to enable passengers to feel that they have had value for money. Leisure-focused airlines are not really that different in attitude. Short-haul airlines, and in particular the low-cost carriers, provide very basic comfort levels.

Seat pitch for Economy-class tends to be in the region of 32–34 inches, although Singapore Airlines is the exception in offering 37 inches for their routes to New York and Los Angeles.

Of course, there are other differentiators, such as the level of service provided. However, long-haul Economy-class passengers expect a degree of lumbar support, an individual IFE system and adequate baggage stowage space. A major difference between long-haul and short-haul is, of course, that passengers will need to sleep in the former type of airline, so little things like winged headrests are important.

One way in which airlines can promote both comfort and economic efficiency is in the use of 'slim' seats. There are many types of slim seats, such as the Recaro 3510, the B/E Aerospace Spectrum, the Weber 5700 and the Sicma Oxygen, for example. The principle is that these seats have a thinner backrest structure, enabling the dual advantage of closer seat pitch with retaining greater shin and knee clearance.

Among the more interesting propositions that have come to the fore is the idea of a cinema-type fold-up aisle seat in Economy-class. Clearly, the intent would be to speed up boarding and avoid having the aisle blocked while passengers load baggage into the overhead stowage. This is an example of how seat designers are tuning-in to the need of the low-cost carriers.

The Premium Economy-class cabin   Invented by Virgin Atlantic Airways, this is often referred to as the 'fourth class' in the aircraft, offering greater comfort to Economy-class passengers who are willing to pay either a premium, or else a full fare. Sometimes, this class of travel can be offered to frequent flyers, according to airline strategy. The development of this type of cabin has come about due to the airlines' identification of this middle market between true Economy and Business. The essence of Premium Economy-class is that the costs are substantially lower than those of Business-class but full-fare Economy yields are not so far away from those of the Business-class. This is precisely the type of cabin that appeals to the independent market segment that we identified in Chapter 3 – in other words, small businesses and self-employed entrepreneurs who have less negotiating power for discounts.

In fact the concept of Premium Economy has been around for a long time. SAS Scandinavian Airlines had 'Euroclass' as long ago as 1981. 10 years later EVA Air launched 'Evergreen Deluxe', but it was Virgin Atlantic Airways that launched the modern version of Premium Economy in 1992. From 1999 onwards, we saw many airlines venture into this segment, such as Garuda, United Airlines, British Airways, British Midland, China Southern, All Nippon Airways and China Eastern, for example.

These products included innovations such as wider seats, improved seat pitch (up to 37 inches in the case of SAS Scandinavian Airlines, and 38 inches for British Airways' World Traveller Plus) with better recline, laptop power ports, personal video, plus enhanced service features, such as better catering and separate check-in.

The Business-class cabin   The first true Business-class was in 1978 when Pan Am introduced 'Clipper Class'. Other innovators through the 1980s were Qantas, Virgin Atlantic Airways and, significantly, British Airways, who pushed the boundaries many times with 'Club World', cradle seats, and lie-flat seating in 2000. The latter innovation involved the creation of both forward and rearward facing seats, configured in a 2-4-2 lateral arrangement on the main deck of the 747-400 and 777. The ability of the fuselage to accommodate such arrangements is obviously critical.

Some airlines that have abandoned or no longer offer First-class tend to configure their Business-class seats in a 2-2-2 lateral arrangement, such as Continental Airlines, Virgin Atlantic Airways or Air Canada. Others, such as Qantas, All Nippon Airways and Japan Airlines, preferred to retain a 2-3-2 lateral configuration. Again, there are implications in terms of overall seat count.

One of the most significant innovations of recent years has been the introduction of cocoon seats. These have become virtually a standard in the arms race. The certainty of a lie-flat bed incorporating a reclining system which does not interfere with the space of the passenger seated behind, has been an enormous success and has essentially raised expectations in this market segment. Naturally, there is a seating density issue, which the seating manufacturers have resolved by proposing 'herring bone', or partially side facing, seating, as preferred by Virgin Atlantic Airways, for example. Incorporating cocoon seats in Business-class will typically burn up around 60 inches in seat pitch. One way of clawing back some precious inches is to provide 'flat', but not necessarily horizontal, lie-flat beds. In this case, selection of the cushion material becomes quite important, unless the airline is prepared to receive complaints from passengers who slide down a slippery surface and end up in a heap on the floor. Seat manufacturers are also mindful that the natural three-degree inclined position of an aircraft in flight means that an angle of 177 degrees is sufficient to achieve a fully-horizontal position.

Cocoon seats are not to every airline's taste. For example, LAN Airlines prefers to save weight by having lighter seats in order to be assured of carrying high-value cargo.

The First-class cabin   There is no doubt that the First-class market is on the decline, as we saw in Chapter 3. Retention of this product is largely an individual policy issue for an airline. First-class is a lucrative market, but there is a huge cost in providing the space and, critically, the service standards expected. Typically, load factors in this cabin are very low, perhaps less than 40% on an average basis, as it is deemed inappropriate not to be able to offer sufficient capacity at all times for these passengers. Hence, there is virtually no spill being generated. It should be borne in mind that some First-class passengers are non-paying, such as government and airline personnel, and this carries a risk of alienating the well-heeled market that does actually pay for First-class.

Airlines have a difficult problem in avoiding cannibalisation of their First-class market in the face of a much-improved Business-class product. However, the product is definitely superior, with the appearance of personal cocoons and mini-suites, and lateral seating configurations of either 1-2-1, or 2-2. Unlike Business-class, First-class lie-flat beds are even flatter, always achieving a 180 degrees horizontal position. Seat pitches can easily exceed 80 inches, with Lufthansa offering 85 inches and British Airways 89 inches.

One – two – three – four classes   The tendency is for short-haul airlines to deliver an effective single-class cabin, with a cabin divider allowing flight-by-flight designation of Business and Economy-class seating. The great advantage is that the cabin configuration is simplified and the creation of the Business-class is limited to purely product issues. These may include, for example, the level of service and whether or not a centre seat in a block of three is deliberately unsold to give the illusion of more space. We can thank the low-cost carriers for democratising the short-haul product.

Two-class operation is becoming very much the norm in many long-haul markets, especially where First-class has been eliminated. Even Singapore Airlines has a two-class configuration (actually Premium Economy and Economy) for its non-stop services to Los Angeles and New York. Airlines that have eliminated First-class have reconciled the loss of this intrinsically lucrative market in exchange for a gain in overall efficiency. Sometimes the sacrifice in terms of space is just not worth it.

Three-class airlines are either the traditional First, Business and Economy type, or else the new breed of Business, Premium Economy and Economy type. Indeed, one can say that in terms of overall levels of comfort the wheel has turned full circle. The emergence of Premium Economy is certainly seen as contributing to a better product mix as the incremental costs are mostly related to space rather than enhancing the service levels.

Finally, four-class airlines have the complication of trying to be all things to all men. Retention of First-class involves a sacrifice of space for a small, sometimes unpredictable, market that requires higher costs and very low load factors. Those privileged few who are lucky enough to enter this cosseted and cosy world are a dying breed.

Multiple-class aircraft mean that airlines are often faced with the tricky issue of how to deal with demand spill between the classes. It is oh-so-convenient to provide upgrades for frequent-flyer card holders. The main issue is that this creates an unreasonable expectation in the minds of passengers that may become difficult to reverse.

Another issue is that multiple classes may not work on every route of a network. Although segmenting the market may seem logical, it does carry high costs due to complexities in revenue management, sales and marketing and product communication and strategy. It is tempting to individually tailor differently-configured aircraft to individual markets. This procedure has been adopted by airlines such as British Airways, South African Airways and Qantas, for example. The downside is that it becomes more complex to interchange aircraft between markets and everything from passenger seat assignment to maintenance becomes a more complicated issue. One way of dealing with aircraft substitutions is to synchronise seat row numbers between aircraft types. Proponents of this include EgyptAir and United Airlines, for example. In the case of United Airlines, many aircraft types are configured differently. For example, the 777-200 fleet comprises no less than three separate configurations, of both two and three classes, depending on whether the aircraft are assigned to trans-Pacific, trans-Atlantic, transcontinental, inter-hub or Hawaii routes.

Monuments and stowage   The amount of galley space, the number of lavatories and the volume of stowage is largely a function of the degree of amenity the airline considers appropriate for both its brand as well as the market segment it is serving.

There is a tendency to eliminate the central overhead stowage compartment in the Business-class cabin in order to create better impression of space. Clearly, this is essential in First-class, where the lower seating density does not create a particular problem of lack of stowage. However, the greater seating density in Business-class, even with generous seat pitch, does create difficulty as anyone who has travelled in a full cabin will testify. There is an obvious weight implication in fitting a central overhead stowage, too.

As we discussed in Chapter 2, some airlines prefer to locate passenger facilities of this nature in the underfloor hold. Such a preference is not limited to a particular type of airline. Both Lufthansa and Airtours, for example, favour underfloor passenger facilities. Much depends upon the need, or otherwise, of space for cargo, as well as one's view on security arrangements.

Airlines will definitely look for a degree of flexibility in the location of their monuments. It is thus essential to be able to relocate galleys and toilets easily and rapidly in the event of a cabin reconfiguration. This could occur either when an aircraft is changing operators or perhaps when a seasonal configuration change is called for. Also, it is helpful, especially for long-haul operations, that galley areas can be turned into social areas during flight. Similarly, the conversion of a pair of lavatories into a single handicapped lavatory is sometimes desirable.

Passenger connectivity   One significant product development of recent years, and a perfect illustration of the arms race, has concerned the widespread provision of individual In-Flight Entertainment (IFE) systems. In fact, IFE is not a new idea, as live musicians were a feature of air transport as long ago as the 1930s. In the 1940s, Pan Am projected films on their North Atlantic services.

In today's aircraft, drop-down monitors and a single-projection screen per cabin have almost completely given way to seat-back displays with a wealth of entertainment opportunities – no matter what the cabin. Passengers today are swamped with a veritable deluge of choice. The 'me-too' philosophy has seen airlines almost fall over themselves to steal a march on their competitors. However, the provision of increasingly sophisticated IFE systems carries a high price. Installation costs of a state-of-the-art integrated IFE package can cost up to $4 million, representing an average 2% of the price of an aircraft.

Among the functionality now considered commonplace is Audio Video on Demand (AVOD), email and SMS via an on-board server with Internet access, satellite television, and landscape cameras. Usage of personal cell phones is just around the corner and a debate is certain to rage over telephone etiquette and whether there should be dedicated (and therefore expensive) areas for passengers to use their cell phones.

There are economic implications for the operation of aircraft as a result of these developments. More passenger amenity means more weight, especially for a top-of-the-market IFE system, which could add several extra tonnes to the weight of the aircraft. A heavier aircraft burns more fuel and increases operating costs. Designers are focused on keeping weight growth under control. For example, the eventual replacement of pre-recorded announcement systems with a simple smart-card saves both weight and wiring. Plasma screens, although available today in the High Street, have yet to find their way on board the aircraft due partly to weight and also to certification issues concerning the potential production of electrical and mechanical interference.

IFE systems absorb more maintenance time, with up to 30 man-hours per week required to undertake complete checks on some wide-bodied aircraft, depending upon the complexity of the installation. As the strategic airline alliances move ever closer to their goal of providing a seamless service throughout their joint networks, they will need to devote more time to integrating their IFE systems. Joint specification of IFE hardware becomes a necessity.

Using new areas of the fuselage   One way to maintain reasonable unit costs of operation, yet benefit from offering a better standard of comfort, is to use areas other than the main deck for various facilities. One obvious solution is to use the underfloor space for toilets and galleys, the so-called 'monuments'. Airtours installed lower-deck toilets in their A330s, resulting in a net improvement in nine seats on the main deck of the aircraft. For example, the lower-deck lavatory option in the A330-200 involves the surrender of two pallet positions and the net gain of nine seats, after taking into consideration the additional seats in the space of five lavatory positions on the main deck and the loss of space for the installation of the stairs. The economic equation should balance the potential revenue of nine additional seats against the loss of cargo revenue. Two cargo pallets equates to 814 cubic feet of available volume, which translates into 3.7 tonnes of payload at a density of 10lb/cu ft. There is also a weight implication of the installation to be considered as well.

Underfloor galleys have also been used in 747s. The 747-8 promises some interesting use of the space over the main deck for passenger bunks, and crew rest options exist in the overhead space in the 777.

New long-haul aircraft designs provide opportunities for brand-conscious airlines to implement quite interesting product features in order to push ahead in the arms race. Reception areas, shower facilities, shops, exercise areas are just some of the innovations that are achievable. Some of these will exercise the minds of the regulators in terms of certification and procedures. All of these will have an economic implication in terms of higher weight and loss of seats.

Some product innovations are implemented with deliberately less thought to economics where the brand has overwhelming importance. One European airline took so much pride in its brand that, when it came to decisions to introduce a bar on their aircraft, cold calculations on the impact on aircraft weight and fuel burn took a back seat.

The fleet planner must be fully aware of how the airline projects itself through its brand. This is especially important because the design of the cabin establishes the personality of the airline. Complex brands need more complex and therefore expensive solutions. Communication of the brand within the cabin covers areas such as uniforms, cabin crew behaviour and service style, menus and entertainment content. The fleet planner must filter out elements that are not directly related to the aircraft itself.

The arms race is definitely present in the long-haul market, but much less so in short-haul markets. This can be explained by the fact that the low-cost carriers have re-educated the market. Passengers do not expect the same frills as they would on long-haul flights. Indeed, passengers might even be disturbed to find an opulent layout on a low-cost carrier's aircraft. The whole point is that you travel on these airlines because they are budget carriers and you expect them to have a frugal cabin and service level! Changing expectations will clearly feed through into how we determine the cabin, so it is quite likely that we shall see a clear division between the cabin being a product differentiator in long-haul markets and the aircraft being seen more and more as a pure commodity in short-haul markets.

Cargo Requirements

We should not omit an analysis of cargo requirements. Cargo markets respond in very different ways to passenger markets. It is uni-directional in nature and is guided by rational decision-makers, unlike fickle and unpredictable passengers. The opportunities are significant, as airfreight is growing at a faster rate than passenger demand. Half of the world's cargo traffic is carried in the underfloor of passenger aircraft, although it is likely that this proportion will decline over time.

Scheduled passenger airlines also pursuing a cargo strategy need to compare the underfloor loading capability of aircraft. Changing the mix of pallet and container positions will alter the overall payload of different aircraft types. Ideal flexibility is where both pallets and containers can be loaded in both the forward and aft cargo compartments.

Combi operation, whereby both cargo and passengers are carried on the main deck, have fallen out of favour in recent years, partly due to stringent regulatory requirements as a result of accidents, and partly because the needs of passengers and cargo differ to such a degree that it is not easy to combine the two payload sets. Also, combi aircraft carry a weight penalty compared to pure passenger aircraft as they are burdened with the installation of a large cargo door on the main deck.

Pure freighters have seen a resurgence, with both new models and conversions of former passenger aircraft on offer. With cargo forecast to grow at rates in excess of the passenger market, a ready market for freight conversions of elderly passenger aircraft is assured.

Defining the Aircraft Configuration

Airworthiness Requirements

There are two principal bodies that regulate the design and operation of aircraft. The US Federal Aviation Authority (FAA) administers aircraft design through what is termed FAR Part 25, with aircraft operation being controlled through FAR Part 121. The corresponding European body is the European Aviation Safety Agency (EASA), established in 2003 with a goal to progressively take over from the joint Airworthiness Authorities' representation of European states that have agreed to develop and implement common safety standards and procedures. Aircraft design criteria are contained in JAR25 and the operational requirements in JAR-OPS-1. Aircraft design criteria are applicable for the entire life of the aircraft programme whereas operational criteria are subject to evolution over time. Despite attempts at harmonisation, there are still differences between the US and European systems.

There are some countries that have produced their own regulatory requirements, such as Australia and Canada. It is always the airworthiness authority of the country in which the aircraft is being operated which determines the rules to be followed. In the majority of cases, local authorities adopt either the FAA or JAA standards.

Defining the Aircraft Cabin

Determining the number of seats is driven by two needs. One is to comply with the airworthiness regulations of the jurisdiction under which the aircraft will be operated, and the other is commercial.

The fleet planner needs to work with a diagram known in the business as the 'LOPA', or, layout of passenger accommodation. This is also known colloquially as the 'fish' and more officially as the Cabin Layout Drawing. It may be supplemented by the Emergency Drawing, which provides information on the location and quantity of emergency equipment. Figure 4.7 shows an example of a LOPA for an A330-200.

In this example, a lower-deck modular crew rest is included. The seating in the Business First cabin shows cocoon seats, which can be easily identified as there is no recline provisioned for the seats which abut door positions, the stowage or partitions. The A330-200 depicted has seating for 259 passengers. However, significant variations may be found. For example, Swiss operates the same type with 197 seats in a three-class configuration, whereas Monarch Airlines flies this aircraft type with no less than 374 seats in a two-class configuration. We shall return to the layout in Figure 4.7 when we will assess the weight and payload of this aircraft in Chapter 5.

Regulations and the maximum number of seats   This issue is directly linked to the number, size and location of the emergency exits installed. Each pair of exits of a particular type can accommodate a certain maximum number of passengers and

30 Business First at 60 inches pitch   229 Economy at 33 inches pitch

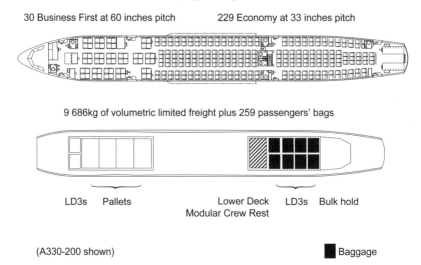

9 686kg of volumetric limited freight plus 259 passengers' bags

LD3s   Pallets   Lower Deck   LD3s   Bulk hold
                                 Modular Crew Rest

(A330-200 shown)   ■ Baggage

Figure 4.7   Typical Cabin and Underfloor Configuration

crew with a presumption that, in an emergency, the doors on only one side of the fuselage would be functioning. There are also rules that determine the evacuation exit path and whether or not space must be provided adjacent to the door for cabin crew to assist in the evacuation. Table 4.1 summarises the general requirements.

In addition to specifications concerning emergency evacuation, there are also requirements concerning the overall geometry and layout of the seats and aisle. The minimum main aisle width should be not less than 15 inches up to the height of the armrest and not less than 20 inches above that. There is a minimum distance required of 35 inches between the 'seat reference point', which is a point for measurement at the base of the seat back, and a bulkhead. Although the minimum seat pitch permissible is 28 inches, it may be the case that in high-density layouts with a large number of seats abreast, the floor loading may require that the minimum pitch be higher.

Accommodation for the crew   Cabin crews are also subject to regulatory limitations. The minimum number of cabin crew should be one per 50 passengers or else the number required for successful evacuation according to a demonstration or analysis, whichever is the higher. In practice, airlines significantly exceed the minimum requirements in the interests of providing quality service, especially for premium passengers. A typical value for a long-haul First-class operation would be one cabin crew member per six passengers. Seating for cabin crew, termed attendant seats in the LOPA, are subject to stringent regulation concerning their design and location within the cabin. It is especially important that cabin crew have vision of the cabin when seated.

As the range capability of aircraft has increased, so has the requirement to provide crew rest facilities, not only for the cabin crew but for the flight crew as well. However, there are no regulatory requirements to provide a rest area for the cabin crew. Underfloor cabin crew rest zones have been designed into the A340,

Table 4.1   Door Exit Limits

Exit type factor	Minimum opening space (width vs height)	Maximum no of passengers per pair of doors	Minimum exit access path	Assist space
Type A	42" x 72"	110	36"	Both sides
Type B	32" x 72"	75 (only in FAR25)	36"	Both sides
Type C	30" x 48"	55 (only in FAR25)	20"	Both sides
Type I	24" x 48"	45	20"	One side
Type II	20" x 44"	40	20"	One side
Type III overwing	20" x 36"	35	6" to 20"	No
Type III	20" x 36"	35	6" to 20"	Adjacent to aisle

whereas Boeing opted for the overhead space in the 777. The most popular form of underfloor cabin crew rest area is the removable type. The module can be extracted from the underfloor hold in around 50 minutes and takes the place of a standard 96 inch by 125 inch pallet. The mobile crew rest for the A340 can accommodate up to seven bunks, whereas the fixed module designed for the bulk hold can accommodate a maximum of 12 bunks. Flight crew rest areas tend to be located adjacent to the flight deck, a solution understandably preferred by pilots, even though they eat into the passenger-carrying zone in some aircraft designs. sometimes a lavatory should be dedicated to crew needs.

Commercial importance of the seat count   Apart from regulatory constraints the number of seats in the fuselage is of paramount importance because the economics will be largely dimensioned as a result. Perennial battles are fought between the manufacturers to ensure that comparative aircraft are configured on a level playing field. With different fuselage cross sections and different philosophies in terms of the number of seats abreast this can become a difficult area of decision. Where unit cost comparisons are made, it is important to understand the interior configuration ground rules if distortions are to be avoided. Once the performance and economics have been calculated for the routes, and the investment appraisal has been completed, it becomes very difficult to go back and change the interior configuration.

Sometimes it is possible to specify precisely the number of seats that the manufacturer should propose in his airframe. Perhaps the market studies have suggested an optimum value. More often than not, it is the geometry of the fuselage that suggests a value. The layout of the aircraft is driven not just by the number of seats abreast, but also by the number of physical classes, their position within the aircraft and the seat pitch. As we have just seen, regulatory constraints for aircraft evacuation determine the maximum number of passengers that can be accommodated between doors.

It is quite common for an airline to precisely specify the number of premium seats and leave it to the manufacturer to fill up the remainder of the space with Economy-class seats. As we have seen in our analysis of market segmentation, it is the premium classes which must be very closely tailored to expected demand and we should now be applying our knowledge of demand spill to ensure that we have the optimum balance between expected load factor, spill and cabin size.

We need to accept that the efficient use of the cabin volume differs dramatically throughout the aircraft. The amount of space occupied by a First-class seat may be more than six times that of an Economy-class passenger. Yet the yield difference may be less than this. Bringing the equation into balance is helped because load factors are higher in the lower yield parts of the cabin. One way of overcoming this problem is to compare aircraft on the basis of the maximum number of seats that can be accommodated in the fuselage and a minimum standard for galleys and toilets. The highest number of seats is the regulatory evacuation limit. Significant differences exist between actual seat counts and the maximum allowable seat count. For example, an A340-500 in typical three-class configuration could accommodate around 313 seats. This expands to a maximum emergency evacuation-limited count of 375, depending upon the emergency exit configuration. However, Singapore Airlines configures this aircraft type with only 181 seats, driven by a combination of market need and range limitation. Although a high-density layout on such a long-range aircraft would be unrealistic, it would give a more homogenous basis for comparing the aircraft. A counter-argument is that this method removes from the equation the advantages that certain fuselage cross-sections have in premium classes.

Reconfiguring aircraft on a regular basis to fine-tune the capacity to market demand is time-consuming and damaging to productivity. However, modern design allows this, even on long-haul aircraft. Lufthansa's A340-600s incorporate one row of Business-class seating aft of Door 2. However, these seats can be substituted with two rows of Economy-class seating when the need arises. The reconfiguration takes only three hours.

One innovative method of varying the seat count without physically changing the installation is through either convertible or variable geometry seating. The latter involves a mechanism to change the width of a grouping of seats. Thus, a triple seat can be easily turned into a double, either with the aid of an electric motor or else a mechanical track. Variable geometry seats are very popular with European airlines, but have yet to become established elsewhere. A more classical way of changing the configuration without changing the configuration is to use movable dividers. This is a useful method of quickly adapting the Business-class and Economy-class configuration according to actual sales.

Catering requirements   Apart from the overall number of seats in the aircraft, the ground rules should also specify the catering requirements in terms of the number of galleys and trolleys. Short-haul aircraft tend to be less problematic than long-haul aircraft in this regard. For long-haul operations there is often a more marked difference in service standards between the premium and low-yield cabins. It is important to determine the number of hot meals likely to be served, for example, and then specify the number of trolleys and galley space needed to fulfil the service requirements.

There are broadly two main types of galleys in world-wide operation; ATLAS, which is the most popular, and KSSU. A distinction must be made between dry galleys which, as the name suggests, are without a water connection, and wet galleys. Chilled galleys require more space to accommodate the air chill system. The usual method of specifying catering space is to determine the number of trays per passenger, including waste bins. Typical values for long-haul flights would be 14 per First-class, and three per Economy-class passenger. For short-haul this might reduce to three per First-class and one-and-a-half per Economy-class passenger.

Toilets   This is mostly a commercial issue, although it is mandatory to provide facilities for the handicapped in the US. Clearly, these facilities are more important on inter-continental routes, which explains why the ratio for single-class short-haul aircraft is around one toilet per 80 seats, and the ratio for First-class long haul aircraft falls to around one toilet for 10 seats or even less. Underfloor toilets release main deck space for additional seating, without compromising on the ratio of toilets per seat. Another advantage is that the disturbance of other seated passengers is removed.

Cabin flexibility   The location of galley and toilet positions was limited to fixed positions right up until the 1990s. In a response to airline requests for more flexibility in the configuration of these so-called 'monuments', galleys and toilets in long-haul aircraft are now mounted on seat rails giving flexibility down to the nearest inch within the designated zones.

An important aspect of flexibility is the time it might take to undertake a cabin reconfiguration that involves, for example, a conversion from a single-class to a two-class operation. This might be needed where an aircraft might be redeployed in a different market for a season. A medium-sized aircraft conversion might absorb between 50 and 70 total man-hours, depending upon the complexity of the task and whether galleys have to be moved. The actual elapsed time could be as low as between seven and 12 hours.

Lead times   As long as cabin options are selected from the cabin configuration guide, then lead times can be assured between the freeze of the cabin definition and the delivery of the aircraft. These lead times may vary according to aircraft type, but can be expected to be around nine to 12 months.

Towards the standard cabin?   This is an issue that divides opinion. On the one hand there are those who advocate differentiation in the cabin, so that airlines can compete vigorously in terms of product. The entire aircraft interiors industry depends upon this philosophy. On the other hand, there is a school of thought that too much individuality in an aircraft cabin is simply a waste of money and that we should all accept that the golden age of air travel is over and the cabin should be as standard as possible. Low-cost carriers have successfully re-educated the market along these lines. Leasing companies certainly would appreciate a vast simplification in aircraft customisation as they may have to bear the burden of reconfiguration many times during the life of an aircraft.

The debate extends beyond the aircraft cabin. The entire aircraft is affected. The big question to be asked, for example, is whether a design, manufacturing and supply process that results in a choice of around 10 different varieties of waste trolley is actually something that enhances the customer experience.

To illustrate the complexities that can arise with a multitude of cabin configurations, equipment variations and different airframe/engine combinations, at one time in the early 2000s British Airways proudly possessed 12 aircraft types, but no less than 44 sub-types.

One illustration of how easy it is to be trapped concerned British Airways' decision to furnish their 767-300 (along with 747-400) aircraft with Rolls-Royce

RB211-524 engines. The problem was that virtually every other airline of the type favoured either General Electric or Pratt and Whitney engines. Faced with this situation the choice is either to accept that the aircraft would always be very difficult to sell, or else to keep them for their entire life.

It is easy to find arguments to support standardisation. Development and manufacturing cost is reduced thanks to economies of scale, along with production time, complexity and capital cost. Standardisation allows production flexibility, faster response and implementation, an easier procurement strategy, improved opportunities for airlines to resell aircraft, and improved residual value. Also, it becomes easier to monitor and compare key performance indicators.

Proponents of customisation argue that competition is essential in the business to drive efficiency, provide what customers are looking for in specific markets and, critically, to spur product innovation. In other words, wouldn't the world be a boring place if everything looked the same!

If we accept that some form of customisation is a necessary evil, then we must be prepared to deal with it from a fleet planning and operational view. For example, the provision of standby aircraft and stock of spares become complex issues. If an airline is going through a process of change or upgrade, then it becomes confusing for passengers. Iberia took delivery of A340-600 aircraft with a new interior, and then had to retrofit the same standard into the existing A340-300 fleet. Regular travellers were easily disgruntled during this transition as there was no certainty as to which cabin would appear on which service.

This leads onto the question of how an airline deals with a mid-life reconfiguration. Here, we need to consider the issue of downtime cost during the reconfiguration, the actual cost of the reconfiguration, revision of residual value of the aircraft, and any additional documentation or ground equipment or training associated with the new product. Also, any post-delivery customisation is not certifiable with the airframe manufacturer's airworthiness authority.

One could argue that alliances should have a significant role to play here. One goal of a strategic alliance is to achieve a high degree of seamless service between the members. This objective goes against the concept of individual branding, especially as all but the very smallest members of a major alliance would be unlikely to accept that the alliance brand becomes more important than their individual brand. However, where alliance members do operate common aircraft types there are better opportunities to achieve at least a degree of similarity in cabin 'look'. The Star Alliance has achieved agreement for a common configuration of a regional jet. However, finding agreement for a truly common and fully interchangeable aircraft seems as far away as ever.

Principles of Aircraft Specification and Customisation

Customisation is usually addressed after the aircraft has been selected. However, it is as well for the fleet planner to be aware of the process involved as certain elements of the aircraft should be assessed with customisation issues in mind. Also, it is usual for the fleet planner to remain fully involved in the process of customising the aircraft, so it is imperative that we discuss the full picture. To conduct the evaluation, the fleet planner should be armed with documentation from both the manufacturer as well as from within the airline itself. Indeed, the fleet planner has a responsibility for structuring some of the documentation that the manufacturer needs in order to simulate aircraft performance. Let's take a look firstly at the manufacturer's input, and then the airline input.

Aircraft Manufacturer Documentation

The manufacturer can supply all manner of data and information according to their potential customers' requirements. This very much depends upon the degree to which a customer is analytical, whether the data exists already, whether the analysis is being performed by a third party, or whether the type or technology standards are already deployed.

The list of documentation that follows is not exhaustive but gives a fair guide to the basic data with which an airline should be conversant.

Standard specification The 'clean sheet of paper' for an aircraft is the standard specification, bereft of customer choices in terms of cabin layout, seats, galleys, IFE system, avionics or any other specific needs. The standard specification is a contractual document that describes the standard aircraft, giving information concerning the basic geometry of the aircraft, the floor loads, volumetric measures, design weights, design speeds and systems.

It is arranged according to the ATA engineering numbering system which comprises a series of chapters arranged in a common format. For example, Chapters 1–20 give the aircraft dimensions, and Chapters 21–80 describe the major systems of the aircraft, such as the air conditioning, autoflight, communications, electrical power, equipment and furnishings, fire protection, flight controls, fuel, and hydraulic power systems. The ATA system is an industry standard, and is applied to areas as diverse as reliability reporting, maintenance and spares provisioning.

The standard specification includes statements concerning the certification status of the particular variant, with reference to certificating authority standards. The manufacturer must also establish additional certification items, such as a Master Minimum Equipment List (MMEL) with approval from the relevant airworthiness authority.

There will be differences between the description of the aircraft in the standard specification and the aircraft to be actually delivered. The buyer may require changes that are notified through a Request For Change (RFC) procedure, and then answered by the manufacturer as Specification Changes Notices (SCNs). The SCN comprises the wording alterations to the specification of the aircraft, any repercussions on equipment and weight, and information concerning the pricing of the change, latest decision date and on which specific aircraft the change can be implemented. There will be RFC/SCNs in respect of the colour specification for cabin furnishings and materials, as well as the external livery of the aircraft, for example. Upon signature, the SCN becomes a contractual document.

Sometimes, an operator has very specific requirements, and these should be considered as part of the aircraft evaluation process. For example, American Airlines require that the underside of the fuselage skin be polished, rather than painted. Also, operators serving routes over the North Pole have equipment requirements for operating in these extreme environments.

The manufacturer reserves the right to make changes to the standard aircraft either to improve the aircraft or to comply with a Purchase Agreement, which is the most important legal document defining the obligations of the buyer and seller.

The Purchase Agreement   This is the contractual document for the sale of the aircraft, and the holy grail of the aircraft Sales Director and source of many sleepless nights.

The Purchase Agreement, or PA, sets out pricing (including conditions for payment and escalation), delivery conditions, initial spares provisioning, training, all guarantees and warranties, and any other negotiated agreement.

The PA also specifies the Buyer Furnished Equipment (BFE), which must be provided by the buyer for installation in the aircraft. The BFE includes communications and navigation equipment, galleys and seats. There is another category of certificated equipment that is selected by the buyer, but installed by the manufacturer. This is Seller Furnished Equipment (SFE), and includes the air conditioning, autoflight, fuel, electrical and hydraulic systems, plus the brakes, auxiliary power unit and the engines. SFE also includes cabin items such as the overhead stowage, toilets, public address and smoke detection systems. All cabin items are detailed in the cabin configuration guide.

Cabin Configuration Guide   This document describes an uncustomised aircraft, or else a family, in a way that a prospective customer can understand the opportunities that exist to build a customised layout. Through a customisation catalogue, an airline can raise whatever RFCs it requires. The standard aircraft contains the floor panels, cabin lining (including side-wall and ceiling panels), lateral and central overhead stowage units where applicable, passenger and cockpit doors, and a number of systems such as air conditioning, electrical power supply, smoke detection systems, lighting, water and waste systems, and wiring for oxygen systems. The cabin Configuration Guide follows the logic of the ATA numbering system.

System Configuration Guide   This document provides information about the optional aircraft systems, their function and alternative equipment vendors.

Aircraft Technical Description   This document provides details of the aircraft, broken down into general information, the aircraft structure, systems and power plant.

In addition to the above, there may be documents that describe the aircraft systems from an operational point of view and provide more detailed descriptions of specific systems. The following set of documents is highly relevant for the evaluation stage of an aircraft.

The Performance Manual   The basic source of payload and range data for the fleet planner is contained in the Performance Manual. Data are specific for each airframe and engine combination and, for aircraft in production, is based upon flight test results. The document may also reflect expected improvements in performance for future delivery standards. If, for example, an engine manufacturer has in place a performance improvement programme, the airframe manufacturer may choose to reflect these improvements, after auditing the proposals, in his manual. The intention of the Performance Manual is to show the anticipated capability of an aircraft in 'nominal' conditions. The document includes basic information concerning the weights, dimensions, engine thrust levels and operating speeds. The manual describes the performance of the aircraft in take-off, climb, cruise, descent, holding, and landing modes and includes data on the engine-out ceiling and stall speeds.

Maintenance Planning Document   This document contains the manufacturer's recommended maintenance programme for its products. The tasks described are based upon Maintenance Review Board Reports and are regularly updated according to the development of the technical status of the aircraft as well as in-service experience. The Maintenance Planning Document acts as a basis for an operator to establish his own maintenance programme that must be approved by the local certificating authority.

The Maintenance Planning Document defines maintenance tasks appropriate for an aircraft utilisation that is not exceptionally low or high. However, each operator should compile his own maintenance programme in accordance with national requirements. Operators must also be aware of both manufacturers' and vendors' recommendations that may be published through Service Bulletins or information letters. Also, the makers of engines, the auxiliary power unit and other components not made by the airframe manufacturer, must supply their own recommended practices to the operator.

Airport Planning Document   The manufacturer provides this manual to provide guidance for both airport operators and airlines concerning the handling of the basic version of a particular aircraft.

Data includes aircraft dimensions such as ground clearances, details of the passenger cabin and cargo compartment arrangements and door positions. Turning capability and manœuvring characteristics are also shown to ensure that aircraft stands and taxiways can accommodate the aircraft. The handling requirements are also outlined, with flow charts showing the times required for deplaning and enplaning passengers, unloading and loading baggage and cargo, and servicing the aircraft. Finally, information concerning the operation of the engines includes the size and shape of exhaust velocities and noise data.

In addition to the basic documentation described above, the manufacturer also makes available, or else customises, a wealth of other manuals concerning the operation of its products. These would include the Flight Manual, Flight crew Operating Manual (FCOM), and Weight and Balance Manual, for example. If the aircraft is certificated for Extended Twin Operations (ETOPS), then a Configuration, Maintenance and Procedures document would be prepared to cover all ETOPS-specific items.

Now, let's take a look at how the all-important ground rules for the customisation of the aircraft should be provided by the airline for the manufacturer's use.

Airline Documentation

There are two basic forms in which an airline might communicate its requirements to the manufacturer. However, there are no hard and fast rules for the way in which a potential customer defines its needs.

Request for Information (RFI)   This is an informal document to enable the airline to pose questions specific to an aircraft type in which it has an interest.

The RFI might, for example, include questions relating to the aircraft's generic characteristics in terms of design weights, engine thrust ratings, range, cabin configurations and design life.

Perhaps the airline would appreciate the manufacturer's view on how the airline market might evolve. Certainly, the airline would need to know the manufacturer's market position in terms of total orders and options of the type, delivery status, number of operators and order backlog.

The RFI would also probably include questions relating to product development projects, expected lifespan and future delivery positions.

The intention is simply to gather information for internal reflection, rather than to challenge the manufacturer to produce a commercial offer at this stage.

Request For Proposal (RFP)   This document is intended to formally outline airline expectations of the composition of the offer it expects. An RFP is a confidential document as it contains much of strategic importance and potential value to a competitor. The RFP should be submitted to competing manufacturers enabling each to submit their own solutions and proposal by the specified closing date.

Content would typically embrace a summary of the status of the airline's fleet situation, with data concerning the types currently operated, and a broad indication of volume of business it expects to generate. The RFP may specify a requirement for a generic aircraft, such as 'a 300-seater', without indicating a particular aircraft type or variant. It is up to the manufacturer to offer a solution that fits the needs of its potential client.

An RFP will indicate a date by which the manufacturer should submit a proposal, and another date enabling enough time for the airline to conduct an internal assessment.

Commercial terms would be indicated, such as the number of firm and option aircraft sought. The distinction needs to be made as it will affect the number of delivery positions the manufacturer may choose to hold and will also affect pre-delivery payments and refundable deposits, which are connected to firmly ordered aircraft. Projected dates for the entry into service of firm and option aircraft should also be given.

Expectations of pricing structure may also be given in the RFP. These might indicate whether any guarantees are expected and the content of an introductory support package. The RFP might also specify conditions for a fleet management contract for the engines, for example. This is otherwise known as a 'power by the hour' deal which involves the operator making regular payments to the owner of the engines, according to an agreed utilisation plan, in return for the owner undertaking operating costs and general support.

The RFP will also include details concerning the specification of the aircraft. These details might cover areas such as avionics standards, cargo loading systems and fire protection standards.

The assumptions   The third form in which an airline outlines requirements is by a 'ground-rules' document. Ground-rules for the analysis and comparison of aircraft are frequently a subject for much debate at the beginning of an analysis. Unlike an Aircraft Specification or Airport Planning Document, there are plenty of grey areas. Make no mistake about it; fixing the assumptions is one of the most critical areas in the entire fleet planning process.

The conditions under which aircraft are compared do not necessarily reflect the conditions under which they may be operated. Indeed, it is hardly ever the case that an evaluation LOPA, or aircraft interior configuration, reflects the layout of a delivered aircraft. This is mostly because it is just not possible to assess with any precision how the aircraft will be assigned onto a network, which may itself evolve between the evaluation phase and the final delivery. Also, the evaluation team may wish to inject a certain degree of conservatism into the ground rules, especially if the aircraft under study is a brand new type with no history.

The manufacturer, on the other hand, will wish to see study rules which enable him to show his product in its best light and will be reluctant to accept study parameters which unduly penalise the performance or economics of his aircraft. It is very important that the same study conditions are given to competing manufacturers.

The study assumptions may replicate data in the RFP, which is a more formal document. Yet the assumptions may also leave some leeway for the manufacturer to prepare his own studies. Sometimes unsolicited study work from the manufacturer can throw completely new light on the airline's views. Manufacturers' sales brochures are important in that they also consider more qualitative aspects of the aircraft whereas ground rules comprise mostly quantitative elements.

When compiling study assumptions it should be borne in mind that they may need to be applied not only to new aircraft currently under production, but to used aircraft which are already in service, and project aircraft for which the level of data will be reduced.

Confidentiality should always be maintained because the content of aircraft evaluation can be interpreted as an indication of the airline's economic potential, as well as give clues as to which routes may be opened in the future.

Study ground rules are usually sub-divided into five categories: interior configuration, operating weights, systems requirements, performance, and economics and financial analysis.

Interior configuration ground rules   The assumptions should provide information concerning desired seating numbers in the various classes, seat pitch, recline and seat width. Passenger amenities should be specified, such as the number of toilets, number of galleys (including volume and weight), number of closets, attendant seats and arrangements for crew rest.

Operating weight ground rules   Here we would expect to find the derivation of the operational items (which will be covered in detail in Chapter 5), passenger weight (including baggage allowances), preferences for carrying passenger baggage and cargo, plus cargo and fuel densities.

Systems requirements ground rules   It is usual to specify any particular systems needs, arranged by ATA chapter. What is important is to identify any item that may have an effect on the aircraft weight, such as emergency equipment, for example.

Performance ground rules   This is often the largest section in the ground rules document. Here we would typically find instructions for the calculation of the take-off, temperature conditions to be applied, and specific dimensions of airfields to be considered in the analysis, including their ambient conditions. For the *en-route* performance we would find information concerning the flight profile, a list of routes to be analysed along with distances to be used and wind conditions to be applied.

How these data are captured, analysed and presented is the subject of Chapter 5.

Economics and financial analysis ground rules   This is an opportunity for the airline to indicate any specific requirements in how the economic and financial analysis should be conducted. For example, the airline may like to see numbers relating to an average sector block time or average daily utilisation. Also, we may find useful information concerning how the maintenance would be performed. It is important for the manufacturer to build a picture of how much maintenance would be undertaken in-house and how much would be subcontracted, for example. The manufacturer should also be aware of the composition of the Direct Operating Cost breakdown.

Lastly, guidance on the financial analysis is importance, with information on the company discount rate and length of study period being typically needed.

The analysis of these data will be covered in Chapters 6 and 7.

In Summary

In this chapter we have reviewed the aircraft product strategies of the two major manufacturers and addressed the significant issue of the aircraft cabin.

The demand for aircraft is expected to remain strong well into the future, and both major manufacturers of large jets have developed broad product lines to cater for all segments of the market. However, there is considerable difference of opinion between Airbus and Boeing on how future demand is likely to be accommodated. Airbus believes that markets will continue to fragment but a need will arise for very large aircraft to serve the densest routes. Boeing sees a larger proportion of deliveries over the next 20 years comprising single-aisle aircraft.

Commonality of technology has become increasingly important, owing to the flexibility it offers and the potential for operating cost savings, particularly in spares and crew training.

We have also reviewed in this Chapter the principle forms of documentation supplied by manufacturers and discussed the all-important airline study assumptions, with particular reference to the significance of seat-count.

At this stage in the fleet planning process we have now reviewed the market, the aircraft types on offer and airline product needs. It is now time to embark on the performance analysis of those types chosen for evaluation.

Chapter 5

Aircraft Performance

Why is Aircraft Performance so Important?

In the days when fleet planning was driven by mostly technical considerations, aircraft performance assumed a key role in the evaluation. There were two basic reasons for this. Firstly, the subtleties of marketing were little known, yet alone applied, in the early jet age. The airlines were not as competitive as they are today as schedules and fares were under the control of the International Air Transport Association (IATA). Secondly, aircraft design was not as advanced as it is today and there were significant limits to what could be achieved in terms of range and payload.

Today, competitive pressures and marketing subtleties have consigned the technical side of aircraft evaluation to a supporting rather than a lead role. Furthermore, dramatic advances in the range capability of aircraft have gradually opened up a vast array of city-pair opportunities. Yet aircraft performance remains significant because there will always be limits to what an aircraft can achieve in terms of payload or range. These limitations frequently form the basis of guarantees which manufacturers are asked to provide.

The underlying purpose of the study of performance is thus to optimise the payload and range abilities of an aircraft according to a set of physical and ambient limitations. The physical limitations concern the configuration of the aircraft and the characteristics of the runways from which it takes off and lands. The ambient limitations concern operational elements such as temperature, wind and airfield elevation.

In this Chapter we will examine the weight build-up of an aircraft, the principal issues concerning aircraft performance comparisons, and how to compose a performance evaluation for a set of routes in the context of fleet planning. This will entail examination of both airfield performance and *en-route* performance. The weight build-up and performance results are a consequence of the assumptions laid down by the evaluators. These assumptions must take into account not only the technical definition of the aircraft and expected loads, but also those all-important ambient conditions prevailing for each of the routes being studied.

Apart from gaining valuable knowledge about the aircraft capability, the performance analysis provides us with payload and fuel burn data that we need for the economic analysis.

Some Background on Aircraft Certification Practice

A word or two on how aircraft are certificated will give us a useful background when dealing with the definition of aircraft weights and performance.

There are two types of certification, the first being a domestic certification under the jurisdiction of the state of design and manufacture. For example, in the case of Boeing this is covered by the US Federal Aviation Regulations (FARs) of the Federal Aviation Administration, and for Airbus has historically been under the Joint Aviation Requirements (JARs) of the Joint Airworthiness Authorities (JAA). However, the creation of the new European Aviation Safety Agency (EASA) in 2003 was the first step in consolidating certification, regulation and safety affairs under a new single body. The A380 was the first aircraft certified by EASA. The large transport aircraft design regulations are contained in FAR Part 25, shortened to FAR 25, and the 'mirror' European version is called JAR 25. There is an on-going programme of harmonisation of the regulatory differences between the US and European systems in order to reduce costly design changes that bring no enhancement in safety. However, the process has been long and erratic, with both sides failing to agree on some key issues. One of the problems concerns how to manage a single legal entity with powers to act on behalf of all members.

The second type of certification involves the validation of the basic aircraft certification by the local, or national, authority of the country in which the aircraft will be operated. In Europe a huge number of so-called national variants required by the many individual states were removed upon the creation of JAR 25.

A major step in the implementation of a European common operating regulation came in 1995 with the adoption by the JAA of the first part of JAR-OPS, or air operations requirements. These requirements, which are applicable throughout the JAA member states, contain regulations on operational procedures, performance, aircraft weight and balance, communications and navigation, flight and duty times and security. Implementation and enforcement of JAR-OPS are in the hands of individual JAR countries that retain autonomy in all regulatory matters.

Aircraft Weight Build-Up

Defining the Initial Weight

Although laws of aerodynamics dimension aircraft performance, operational weights are determined by design objectives. We should therefore start by defining the main categories and then building the weight of the aircraft.

Many of these weights must be approved by the certificating authorities and are quoted in the Flight Manual and Weight and Balance Manual for the particular aircraft. We must distinguish between those weights that are certificated and those that are essential to achieve a certain payload uplift from an airfield and reach a certain destination. Required weights are specific to a mission; certificated weights are maxima that are dimensioned by structural limits.

Figure 5.1 shows how an aircraft weight is built-up, starting from the weight of the structure, adding elements to make the aircraft operational, then the payload, then the fuel, and finishing a mission with a final weight according to how much fuel has been burned.

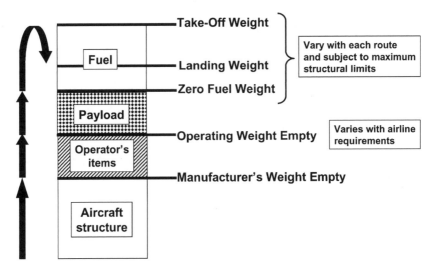

Figure 5.1   Building Up the Aircraft Weight

Manufacturer's Weight Empty (MWE, or sometimes MEW)   This is the weight of the aircraft as it has been built by the manufacturer, without the elements necessary for revenue earning payload to be carried. Therefore, the MWE does not include (usually) the seats, galleys, pallets or containers. Nor does the MWE include fuel or payload or other items that are necessary for the actual operation of the aircraft. The MWE sub-divides into two portions, the manufacturer's MWE and the airline MWE. The latter will include the added weight of any changes that the airline specifies over and above the standard specification aircraft weight. Naturally, at the beginning of an evaluation the airline would probably not have made detailed decisions as to what changes it would like, so it is usual to add a percentage, such as 1–2%, to the manufacturer's MWE.

Operator's items   Here we must add those elements that are necessary for an aircraft to be actually operated in service. The list of items is fairly standard and comprises the following:

- Unusable fuel (i.e. fuel which remains in pipes which cannot normally be purged from the fuel system)
- Oil for engines and APU
- Water for galleys and toilets
- Fluids for toilets
- Aircraft documents and tool kit
- Passenger seats and life jackets (if required)
- Galley structure and fixed equipment
- Passenger service items (such as all catering, pillows, blankets and give-aways)
- Emergency equipment (such as slide rafts and first-aid kits)
- Crew and crew baggage.

Most of the decisions concerning the make-up of the operator's items are made by the airline rather than the manufacturer. For example, the definition of passenger service items is purely a brand and airline product issue. The number of meals, snacks and drinks to be served, plus in-flight sales of duty-free products, will all impinge on both space and weight.

Operating Weight Empty (OWE, or sometimes OEW)   This is the weight of the aircraft prepared for service and is composed of the MWE plus the operator's items.

The OWE does not include any payload or fuel. Sometimes the weight of pallets and containers are considered as payload, in which case they are deleted from the list of operator's items. The level of OWE controls the structural payload of the aircraft.

When building the weight for study purposes it is important to know whether the aircraft is being represented in delivery condition or at some future point in time. As aircraft become older they tend to get heavier, just like human beings. This can be the result of an accumulation of debris in inaccessible areas in aircraft, inaccurate record keeping, plus a string of cabin modifications and Service Bulletins imposed by the manufacturer. All this can mean, for example, that up to 150kg can find its way into the structure of a 747 each year. The value for a single-aisle aircraft would obviously be much lower. This extra weight eats into the usable payload. A logical way of dealing with this is to ask the manufacturer to quote the MWE and OWE at mid-life. During heavy maintenance checks the airline will attempt to remove some unnecessary weight from the aircraft which may have accumulated over time.

Airlines sometimes request an MWE tolerance percentage to be added, in addition to the in-service weight growth value, to reflect variations in actual delivered aircraft from the specification. This may be eliminated if a guarantee were to be provided by the manufacturer.

Maximum Zero-Fuel Weight (MZFW)   It is the weight of the aircraft with payload added, but no fuel in the tanks. The MZFW minus the OWE gives us the 'structural payload'. The MZFW is a certified weight and is limited by the bending moment at the root of the wing.

Maximum Design Take-Off Weight (MDTOW)   This is the maximum certified weight at which the aircraft can take-off, as measured at the brake release point on the runway. Adding taxi-fuel to the MDTOW gives a higher value that is called the Maximum Ramp Weight. The MDTOW can be limited by many factors, including engine thrust, brake energy or tyre speed, for example. Sometimes lower MDTOWs than the theoretical maximum are deliberately certificated for a particular aircraft, so the absolute maximum value for the design is referred to as the MDTOW, and lower values are called more simply the MTOW. Often the latter expression is used in place of MDTOW. The MTOW primarily determines the range of the aircraft.

Once we add fuel to the tanks we can now perform a mission. We shall see in a moment the interaction between the MTOW and the fuel capacity of the tanks. The aircraft must carry sufficient fuel not just to get from A to B but also to attain a diversion airfield if a landing at the destination is not possible. The reserve fuel, which we will look at in more detail later, must also include various allowances to take account of changes in operational conditions (especially winds) between the time of computation and actual departure, which might amount to several hours. These fuel reserves are a crucial part of the performance calculation and determine to a significant degree the payload that can be carried. Once the aircraft arrives at its destination it will land at the appropriate landing weight for the mission. Each aircraft can land up to the maximum landing weight certificated.

Maximum Landing Weight (MLW) The maximum certified weight at which an aircraft can land is determined by the loads that impact on the landing gear. The value consists of the OWE, plus the payload and any reserve or allowance fuel not consumed. Unsurprisingly, the MLW is inferior to the MDTOW but there are some circumstances, such as an emergency just after take-off, where overweight landings are permitted.

Building the Payload

To do this we must already have to hand a number of elements and it is useful to divide them into main deck and lower deck because changes in one area will affect the other.

Main deck payload The cabin layout must have been defined to the point where we at least know the number of seats by class, the number of galleys and toilets and how many attendant seats are necessary. The finer details that have an impact on the weight are less critical at this stage, but we should have an idea of how much additional weight needs to be added to the structure to represent the changes to the standard aircraft.

A weight should be assigned for each passenger plus checked-in and carry-on baggage. Weights used in aircraft evaluation are based upon observed actual values or else specified in regulatory documentation, such as JAR-OPS 1.620. The 'standard mass values' specified by JAR-OPS for aircraft with a seating capacity of 30 seats or more is 84kg per adult for all flights except charters, 76kg for charter operations, and 35kg for children. Alternatively, it is possible to use a male and female weight breakdown of 88kg and 70kg respectively for all flights except charters, and 83kg and 69kg for charters. Baggage values range from 11kg for domestic flights to 15kg for intercontinental flights, although in practice much higher weights are usually recorded.

Operators who wish to use alternative values should gain the approval of their local authority through passenger weight sampling of up to around 2,000 passengers. It is often the case that, for study purposes, aircraft configured with premium classes be assigned a higher weight allowance, to reflect higher permitted baggage allowances. Typical values range from as low as 90kg for charter Economy-class passengers (including baggage) up to as much as 115kg for scheduled First-class passengers. Such a dramatic variation, coupled with the effect of different seating configurations, has an impact on the maximum passenger range for a particular aircraft.

Table 5.1   Weight Breakdown Example – A330-200

MDTOW	233 000 kg
MLW	182 000 kg
MZFW	170 000 kg
MWE	106 530 kg
MWE customised	109 417 kg
Operator's items	17 752 kg
OWE	127 170 kg
Structural payload	42 830 kg
MZFW minus OWE	

Lower deck payload   The underfloor arrangement must be decided, to the extent that we know whether we will be accommodating pallets or containers, or else underfloor passenger facilities. The baggage allowance per passenger has to be fixed, together with the density. The number of containers required for passengers' baggage can either be determined by calculating the volumes, or else, more simply, a certain number of passengers' baggage can be assumed per container. A typical value would be 35 passengers' baggage to fill a LD-3 container (also known as an AKE, the most widely used container for wide-bodied aircraft). A multi-class cabin can sometimes be penalising for the efficient use of the underfloor where, for example, total separation might be required between First-class, Business-class and Economy-class LD-3s. Also, baggage volumes vary according to geographical region, whether the operation is scheduled or charter and whether routes are short and domestic or long and intercontinental.

It should be decided whether freight would be carried exclusively on pallets, or else carried in a mixture of pallets and containers. The densities of freight should also be determined. Aircraft that do not offer any form of underfloor containerisation have the advantage of not bearing the weight of a cargo loading system or the weight of the pallets and containers. However, they also have a significant disadvantage in that loading and unloading is more time-consuming, cargo and baggage are more prone to damage, and loading staff are more prone to injury. These elements are often difficult to quantify but are, nevertheless, very real and should not be ignored in an evaluation.

When loading either the underfloor hold or a container it must be borne in mind that the 'water volume' of space available can never be realistically filled. Inefficiencies in stacking mean that a percentage of the true volume should be assumed. A usual solution to this problem is to take a stowability factor of between 80% and 90% of the volume available.

The bulk hold of the aircraft consists of a much smaller space at the aft of the fuselage and is often reserved for crew baggage, mails and last-minute checked-in baggage. Owing to the generally awkward shape of bulk hold compartments the stowability factor can be anything from 50% to 70%. On some aircraft alternative uses for the bulk hold are possible, such as the provision of underfloor crew rest facilities or else the creation of an additional container position.

The resulting calculation of usable payload will then determine whether the aircraft will be volumetric or else structurally limited according to the conditions laid down.

Table 5.2   Available Volume Example – A330-200

30 Business First-class and 229 Economy-class
Passenger weight: 80 kg   Baggage weight: 25 kg   Baggage volume: 4 ft³
(per passenger)
Underfloor available volume:

	LD-3 containers (10)	*Pallets (4)*	*Bulk (total)*
Volume	158 ft³	407 ft³	695 ft³
Stowability	85%	85%	70%
Usable volume	134 ft³	346 ft³	486 ft³
Cargo capacity	608 kg	1567 kg	2204 kg
Density :	4.53 kg/ft³ or 160 kg/m³		
Tare weight of pallet :	100 kg		
Tare weight of LD3 :	85 kg		
Cargo:	**1216 kg**	**6266 kg**	**2204 kg**

Baggage:	Volumetric cargo (+ tare)	: 10256 kg
LD3s	Volumetric cargo (net)	: 9686 kg
Required: 7.62	Passenger and baggage	: 27195 kg
Used: 8	Volumetric payload (net)	: 36881 kg
	Structural payload	: 42830 kg

In Chapter 4 we saw a sample layout of an A330-200 (Figure 4.7). We shall now proceed to develop the weight and payload for this configuration. Table 5.1 shows a breakdown of the certified weights of the aircraft, plus an example of appropriate customised operating weights. Table 5.2 outlines a typical volumetric calculation. The example has been deliberately constructed to illustrate a volumetric limitation, which is a function of the densities used. In this case, the customisation of the MWE has added almost three tonnes to the manufacturer's specification value and the structural payload, once the operator's items have been added, amounts to 42,830kg. However, owing to the densities of the passenger bags and cargo, the volumetric limitation is only 36,881kg.

Building the Payload-Range

We will now look in more detail at how the weight of an aircraft is built-up with reference to the payload-range envelope. Figure 5.2 shows the traditional three-sided envelope from which we can determine how much payload can be flown over what distance, according to a set of operating conditions.

It is usual to draw a line to represent the maximum passenger range of the aircraft. This weight includes the weight of the passengers and their baggage. All payload above this line would be cargo. There might also be another line which would represent the volumetric limit of the aircraft, should this be more limiting than the structural limit.

The two kink-points in the envelope occur because of a change in the aircraft weight and fuel volume limits. When a full payload can be carried the aircraft is said to be limited by MZFW. At the first kink point, marked 'A', the aircraft is limited by MTOW, and beyond the second kink, point 'B', the limitation is Maximum Fuel Capacity, or

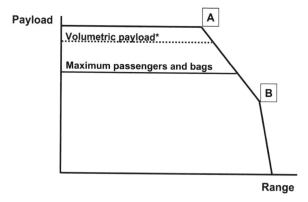

* If volumetric payload inferior to structural payload

Figure 5.2   The Payload-Range Diagram (1)

MFC. In order to appreciate the meaning of these limitations, we need to consider the payload-range diagram together with the various weights we have just defined.

An aircraft OWE means that, by definition, there is no payload and, until fuel is put into the tanks, we are going nowhere. As fuel is gradually added the aircraft has the ability to fly further and further and the take-off weight of the aircraft steadily increases. This situation continues until we reach point A in Figure 5.3, which is the MDTOW. We can never exceed this certificated weight, but it is a characteristic feature of aircraft design that when the MDTOW value is attained, the fuel tanks are not completely full.

In order to continue to fly further, we now need to enforce a trade-off between the weight of more fuel in the tanks and a corresponding sacrifice of payload. This is a straight-line relationship which now continues until the fuel tanks are, indeed, full. The straight-line trade-off is called the MDTOW, or MTOW, limitation and continues up to point B in Figure 5.3. It is always useful to get a feel for the rate of payload sacrifice to gain additional range for a particular aircraft. For a 747-400 one additional nautical mile is worth around 18kg less payload and for an A320 the payload sacrifice would be about 5kg. In the latter case this would mean that one passenger, plus baggage, is equivalent to around 20 nautical miles on the MDTOW slope of the payload-range diagram.

Once we have reached point B on the chart, we can still eke out more range. It is no longer possible to add fuel, because the tanks are full, but we can make the aircraft lighter, thereby improving its range capability, although only marginally. We can reduce the weight of the aircraft by eliminating payload until, theoretically, we have none left and have reduced the weight of the aircraft to that equivalent to the OWE, but with fuel added. This is shown at point C in Figure 5.3. The shape of the payload-range envelope (as shown in Figure 5.2) has now emerged.

It is usual to indicate the range that could be achieved with a full load of passengers and their baggage. This range, along with that achieved with maximum structural payload, is the most often quoted reference to indicate the ability of a particular aircraft.

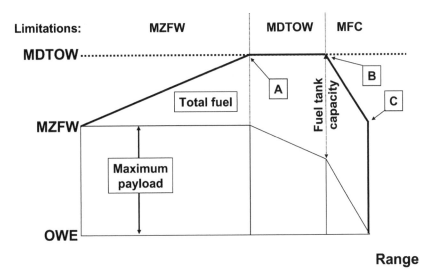

Figure 5.3   The Payload-Range Diagram (2)

How Design Changes Affect the Payload-Range Envelope

It is often the case that one aircraft type can be associated with numerous payload-range shapes according to the variant. We shall now see how each of the three sides of the envelope can be affected by both design as well as by operational practice.

Changing the MZFW or OWE limit   The topside of the curve is limited by the aircraft's Maximum Zero-Fuel Weight (MZFW). If the manufacturer can improve this certificated value by demonstrating the structural integrity of the airframe, then more payload can be made available up to the Maximum Design Take-Off Weight (MDTOW) of the aircraft. The MZFW is a fixed value, whereas the OWE varies according to the weight of the aircraft structure (through the MWE) and the efficiency of the airline's operation in keeping the operator's items under control. When interpreting a payload-range it is helpful to know whether the OWE reflects delivery conditions or the aircraft's mid-life. The effect of a higher OWE on the payload-range curve is shown in Figure 5.4.

Changing the MDTOW limit   The first angled part of the curve is limited by the MDTOW. If this could be increased, then more fuel could be loaded before the trade-off sacrifice of payload has to start. The position of this part of the curve can be affected by either a change to the MDTOW or, in an operational context, by a change to the MTOW.

In the former case, a development in MDTOW by the manufacturer would enable the aircraft to either carry more payload at a given range, or fly further for a given payload, or a mixture of both. As an aircraft design matures it is quite likely that MDTOW enhancements become offered, as the structural integrity of the airframe becomes better known. The effect of a design weight increase is shown in Figure 5.5.

Buying the Big Jets

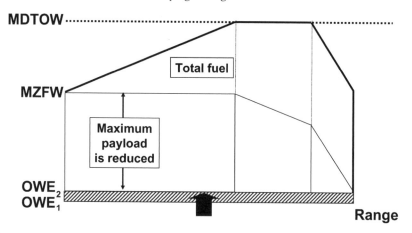

Figure 5.4   Effect of Higher OWE

Very often, design weight increases can be offered with very little additional weight built into the structure itself. If the structure were to be affected, then there would be a marginal reduction in the payload, as a result of raising the MWE. (This is not shown.)

There are sometimes specific operational limitations that mean that the take-off weight which can be used at a particular airfield are less than the basic design of the aircraft. We shall see later how hot-and-high airfield conditions eat into the permissible take-off weight. In such a case, the MDTOW line is driven towards the left of the payload-range diagram, creating the opposite effect to that described above.

Changing the MFC limit   Finally, the steepest part of the curve at the extremity of the envelope is governed by the Maximum Fuel Capacity (MFC). This is the part of the envelope when the tanks are full and extra range is only available by sacrificing

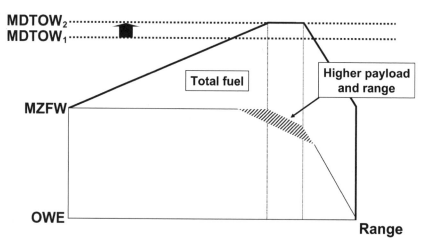

Figure 5.5   Effect of Higher Take-Off Weight

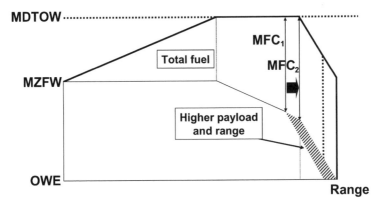

Figure 5.6   Effect of Higher Fuel Capacity

payload and making the aircraft lighter. Increasing the capacity of the fuel tanks delays the point at which the MFC effect takes hold.

The classical way of improving the fuel capacity of an aircraft is to fit additional fuel tanks in the underfloor space. Although range is improved as a result, there are some disadvantages. Firstly, cargo tanks take up space that might otherwise be used for cargo, or even containers for passengers' baggage. Secondly, the weight of the fuel tanks increases the MWE of the aircraft, thereby reducing the available payload. Thirdly, the range improvements are only available where the payload which is carried is at a point on the envelope where the range would otherwise have been limited by MFC. The latter consequence means that it is only realistic to consider underfloor fuel tanks if the required payloads on the longest routes are known to be low and the MDTOW is already quite high, which means that the MFC comes into play where payloads are already quite high. Figure 5.6 shows this effect.

It is often the case that increased fuel volume is accompanied by increased take-off weight in order to 'push' this particular corner of the payload-range diagram to its most effective position. The most common uses for additional cargo tanks are for corporate and VIP aircraft where the payloads are very low and the need to use the underfloor space for cargo is much reduced. The weight of the additional cargo tanks would also marginally increase the OWE. (This is not shown.)

Changing the MLW limit   A Maximum Landing Weight (MLW) operational limit occurs where the distance between the destination airfield and diversion airfield is relatively long, necessitating a high amount of reserve fuel. So, when the aircraft arrives at the normal destination, it is already at its MLW because it must carry enough fuel to fly onto the designated alternate airfield.

It is sometimes, but rarely, the case that the MLW affects the shape of the payload-range envelope. This occurs where the certificated MLW is insufficient. In this case, we need to bear in mind that the fuel loaded is intended partly for the mission itself, partly as reserves in case of a need for holding or a diversion to an alternate airfield, and partly as an allowance which is a percentage of fuel burnt. This latter element varies according to the distance flown. As the allowance is a function of trip fuel, the

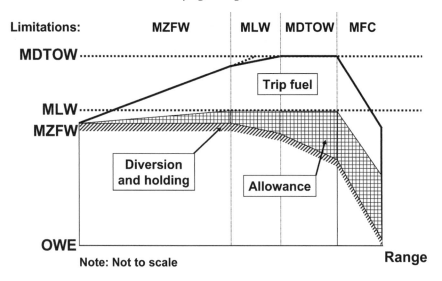

Figure 5.7   Effect of Insufficient Landing Weight

reserves might reach such a value that payload would have to be reduced in order not to breach the MLW limit. This relationship is shown in Figure 5.7.

Interpreting the Payload-Range Envelope

The easiest way of getting a feel for what an aircraft can deliver in terms of performance is to refer to manufacturers' payload-range diagrams. This is especially relevant at the 'rough cut' stage. One can see at a glance if there might be a shortfall in either payload for a certain range, or else range for a certain payload. It is also very useful to see whether the maximum passenger range intersects with the part of the envelope limited by take-off weight or else maximum fuel capacity. In the latter case, there would be a greater potential variation in payload for small changes in range. This means that flexibility in payload planning is reduced compared to the former case, where the trade-off between range and payload is more generous.

Payload-range envelopes are no substitute for a full mission evaluation, as they do not reflect actual operating conditions. Payload-range envelopes are always calculated in 'still-air' conditions, with no wind taken into account. Also, each calculated point along the curve presumes a standard diversion profile, whereas in reality each destination would need to be assigned a specific diversion airfield, the distance of which from the destination determines the fuel reserves to be loaded.

Another problem is that we need to be very clear whether the payload we show on a diagram is the structural or the volumetric payload of the aircraft. Dense freight tends to mean that the full volume of the underfloor holds cannot be fully used, whereas low-density cargo means that the structural loads are not even reached.

Pitfalls of payload-range comparisons   Comparing aircraft payload-ranges is never easy because so much else is hidden. As an example of a pitfall let's consider the case

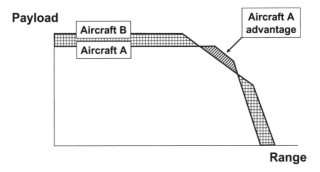

Figure 5.8   Pitfalls of Payload-Range Comparison

in Figure 5.8 where Aircraft A offers a mixture of both higher payload and higher range between ranges at the extremities of the MDTOW limits. What is ignored is the fact that Aircraft B offers a considerably higher payload at shorter ranges and, where payloads are low, a significant range advantage too. Thus, in order to truly conclude which aircraft has the best payload-range envelope, one would have to carefully consider the network on which the aircraft would be operated. In this context, the aircraft with the seemingly better range at maximum passengers may not be the most suitable after all.

It is worth exploring a little deeper this issue of how an aircraft payload-range envelope is fully exploited in reality. Many fleet planners rightly point out that the ability of an aircraft to generate revenue opportunity is linked to its range and payload productivity potential. Range carries a premium in both original price and in future value. Aircraft of any design or type which were early off the production line with lower certificated MDTOW are disadvantaged because their range is limited. We saw in Figures 2.1 and 2.2 that aircraft are rarely used consistently at their maximum range. This reinforces the view that aircraft performance has become much less of an issue than it was in the past. Also, the analysis suggests that it is important to consider the operating efficiencies of even long-range aircraft on short sectors. In fleet planning, it is not the maximum capability of an aircraft which is important, but *where* the aircraft are actually used.

We are now ready to undertake the performance studies. These should always be considered in two parts: the airfield performance and the *en-route* performance. Although we shall not enter into the complex calculations and regulations surrounding aircraft performance, we must nevertheless cover the main issues. It is vital that fleet planners have a good working knowledge of performance principles so that aircraft can be correctly evaluated.

The Airfield Performance Analysis

The weight at which an aircraft can take-off or land at any particular airfield might be constrained by the airfield physical layout, the aircraft design, runway loading limitations, environmental constraints or ambient conditions. Virtually all aspects of these limitations are beyond our direct control, yet all of them can affect the amount of payload that can be carried and the range of the aircraft.

The essence of a take-off calculation is that an engine failure is presumed to occur at a critical point in the take-off run. If the failure occurs up to the critical point the aircraft must be able to stop on the runway, and if the failure occurs beyond the critical point the aircraft must be able to continue the take-off. Airworthiness authorities have laid down minimum values of aircraft path slopes and obstacle clearance. From these, the maximum permissible aircraft weight and associated speeds can be calculated for a set of runway conditions.

We shall now examine the principle limitations which shape the take-off of an aircraft, being the runway, minimum gradient in climb-out (or 'second-segment' limitation), obstacles, tyre speed and brake energy. The take-off weight required must be the lowest after satisfying all of these limitations, subject to an optimisation by adapting the aircraft speeds and flap settings. Finally, we shall look at various parameters that influence the result, such as the wind, runway slope, temperature, runway elevation and pressure altitude.

The aircraft manufacturer is able to supply customised take-off performance for aircraft evaluation purposes and, needless to say, each and every real take-off is also subject to exhaustive calculation.

We should start by setting out some important definitions that concern the speed of an aircraft in the take-off phase of a flight.

Speed Definitions

V_{MCG} Minimum control ground speed from which a sudden failure of the critical engine can be controlled by use of primary flying controls only. The other engine, or engines, remain at take-off power.

V_1 Speed at which the pilot can make a decision, following a failure of the critical engine to either continue take-off or stop the aircraft within the limits of the available take-off or runway length.

V_{EF} The engine failure speed, which occurs before V_1, the difference between the two speeds being recognition time of the failure.

V_{MBE} The maximum brake energy speed to enable the aircraft to come to a complete stop. Energy is dissipated as a function of aircraft weight and the square of the speed. V_{MBE} is also affected by ambient temperature and pressure, runway slope and the wind component.

V_R Speed at which rotation is initiated to reach V_2 at an altitude of 35 feet. It cannot be less than V_1.

V_2 Take-off safety speed reached at the 'screen height' altitude of 35 feet above the runway surface with one engine failed, and which must be maintained for that part of the climb with take-off flaps setting. The lower limits of V_2 are $1.2V_S$ and $1.1V_{MCA}$.

V_S Speed at which lift is lost and the aircraft stalls.

V_{MCA} Minimum flight speed at which an aircraft can be controlled with 5 degrees maximum bank, in case of a failure of the critical engine, the other engine remaining at take-off power with take-off flaps set and gear retracted.

V_{LOF} The speed at which the aircraft wheels lift off the ground, as determined by the take-off weight and flap settings.

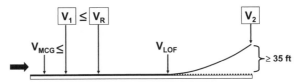

Figure 5.9   Take-Off Speeds

Maximum tyre speed is the maximum speed of the aircraft on the ground, as determined by the tyre strength.

The runway   A number of important definitions exist for runway lengths, and we must distinguish between those distances which are designated as 'available' and those which are 'required' for a specific take-off.

Figure 5.10 shows the Take-Off Run (TOR), Accelerate and Stop Distance (ASD) and Take-Off Distance (TOD). There may or may not be a stopway and clearway present. A stopway can bear the weight of a decelerating aircraft in the event of an abandoned take-off, but is not to be taken into account for the take-off calculation. However, a stopway *is* part of the ASD. The clearway is an additional space beyond the end of the airport-controlled runway, with no obstacles, which can be over-flown before the V_2 speed is attained. It may consist of land or sea.

For a given aircraft weight there is a Take-Off Distance (TOD) *required*, to enable the aircraft to start its roll from brake-release, pass through the decision speed of V_1, lift-off and clear an altitude of 35 feet at the end of the runway at the speed of V_2.

Then, there is the Acceleration and Stop Distance (ASD) *required* which, for a given weight, is the distance necessary to ensure that the aircraft can come to a stand safely on the runway if the take-off is aborted at the most critical point.

Finally, the Take-Off Run (TOR) *required* is the distance from brake-release point up to a point halfway between lift-off to the 35 feet point.

The take-off weight of the aircraft must be such as the TOD, ASD and TOR required must be equivalent to the TOD, ASD and TOR available, often referred to as TODA, ASDA and TORA.

The optimum take-off weight is a function of the decision speed, V_1. As V_1 increases then, for a given weight, the TOD required decreases as the aircraft can take-off with

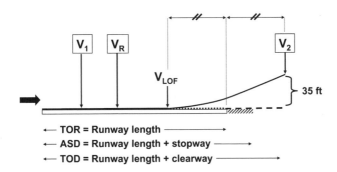

Figure 5.10 Runway Definitions

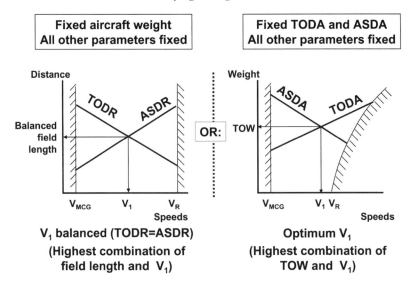

Figure 5.11   Take-Off Decision Speeds

less runway. Conversely, increasing V_1 means that the ASD required increases. The intersection of the two curves gives the balanced field length, as seen in Figure 5.11.

Another way of looking at the problem is to consider that as V_1 increases then more and more take-off weight can be lifted off a given runway length. However, higher speed means that the distance required to stop the aircraft will also increase. Figure 5.11 also shows this complementary relationship.

Runway slope   Anyone who has ever stood at the end of a runway and watched an approaching aircraft appear out of the ground will appreciate that runways are not always totally flat. The average longitudinal runway slope is measured in a percentage that denotes the difference between the maximum and minimum elevation along the runway centre line divided by the runway length. If the overall slope from the runway listed heading is up, a positive (+) sign is used. The regulatory maximum slope is + or − 2%. Figure 5.12 illustrates an example.

Amendment 42   In 1978 the FAA published an amendment to FAR Part 25, known as Amendment 25–42, intended to provide a greater safety margin in the event of an aborted, or rejected take-off. The ruling required that take-off calculations allow for an additional 2 seconds of acceleration after an engine failure at critical point (V_1) before a decision to abort the take-off.

The effect of this ruling was to reduce the maximum permitted take-off weight in order to ensure that the aircraft can stop on the runway. The acceleration-stop distance therefore became more limiting.

Amendment 42 was controversial upon its introduction, as it did not apply to derivative or previously certificated aircraft, thereby distorting comparisons between aircraft of different generations. In 1993, a rule change was proposed, called 'post amendment 42', stating that the 2 seconds of continued acceleration would be

Runway length: 2 600 m

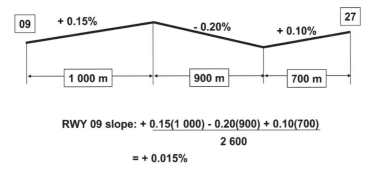

RWY 09 slope: + 0.15(1 000) - 0.20(900) + 0.10(700)

2 600

= + 0.015%

Figure 5.12 Runway Slope

replaced by 2 seconds at constant V_1 speed. Also, the condition of the runway (wet or dry) would be taken into account, and the calculation should be made based on brakes worn to their overhaul limit. These changes were incorporated into the FAR and JAR requirements in 1998 and 2000 respectively.

Line-up allowances   Some airline study rules insist that the manufacturer reduces the length of the runway in order to allow the aircraft to turn and position itself at the end of the runway. The JAR-OPS regulations now include provision for operators to take into account aircraft line-up when computing take-off calculations.

Climb Performance

The next parameter of importance in the take-off calculation is climb performance. We have seen how the calculation is dimensioned around the ability of an aircraft to continue take-off after a failure of an engine at the critical point on the runway. Also, we have referred to the speed V_2, which should be attained at a point of 35 feet above the end of the runway. In fact, there are a series of limitations concerning the climb-out of the aircraft. As always, it is best to start with some definitions.

First segment climb   This starts at 35 feet above the end of the runway and ends when the landing gear retraction cycle is complete. The climb requirements for this initial segment are 0.5% gradient for a four-engine aircraft and 'positive climb' for a twin.

The second segment   This starts when the gear retraction is completed and ends at 400 feet above the take-off runway elevation. Four-engine aircraft must achieve a 3% climb gradient, and 2.4% for twins.

The third segment   This is also known as the acceleration phase and must be accomplished at a minimum altitude of 400 feet. The slats and flaps must be retracted during this phase.

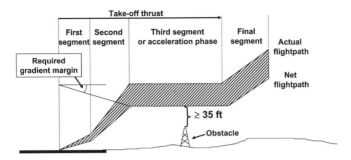

Figure 5.13 Take-Off Flight Path

The final segment   This is also known as the transition phase and continues until at least 1,500 feet above runway elevation. Four-engine aircraft must have a gradient of 1.7% and twins must achieve 1.2%.

The distance between the 35 feet point and the point where final take-off speed is reached is usually about 50km from the end of the take-off distance. Maximum take-off thrust may be used for up to 10 minutes in the case of engine failure.

Obstacles

Perhaps it is stating the obvious, but when an aircraft takes off from a runway it is likely, at some stage, to pass over obstacles in the flight path. These obstacles, whether they are buildings or mountains, may impinge upon the allowable take-off weight which is calculated according to the ambient conditions. If obstacles are present the runway length must be artificially reduced in order to ensure that the aircraft's weight is such that a better climb performance can be achieved to clear the obstacles. There are regulatory margins applied to the actual, or 'gross', path. The resulting 'net path' is computed by reducing the gross path by 1% for aircraft with four engines and 0.8% for twins. Figure 5.13 shows the full take-off flight path, including the effect of the gradient reductions to be applied.

Obstacle locations are usually measured in relation to the start of the take-off run, with the elevation quoted in feet and the distance to the obstacle quoted (confusingly) in metres. Occasionally, the distance might be quoted from the end of the take-off run. If this is the case the manufacturer must be clearly informed so that his calculations take this into account.

How obstacles need to be taken into account for the take-off calculation is a question of regulatory jurisdiction, distance of obstacles from the runway centre-line, whether the take-off is performed with a turn, flying rules and even time of day. Consideration of obstacles has to be in two dimensions; vertical clearance and horizontal deviation from the aircraft track. Figure 5.14 illustrates the principle of vertical clearance, and Figure 5.15 shows the JAR-OPS definition of horizontal clearance. The FAA rule differs.

One of the greatest mysteries in the world of aircraft performance analysis is that it is possible for the location and height of obstacles in the take-off path of an aircraft to differ according to whom you talk.

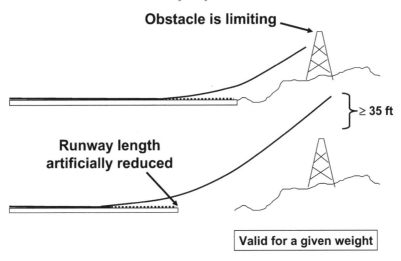

Figure 5.14 Take-Off with Obstacle Clearance

Differences in databases abound, and it is important that both airline and manufacturer agree on which obstacles must be considered for the calculation. The author can recall one bothersome obstacle that was affecting a take-off weight calculation in a performance evaluation. The matter was solved when the airline fleet planning department established that the obstacle was, in fact, a tree. Environmentalists notwithstanding, agreement was quickly reached that any future payload restriction could be easily lifted by the judicious use of a chainsaw!

To recap, we have so far considered the limitations on the take-off of the runway, the margins on climb performance and the effect of obstacles in the take-off path.

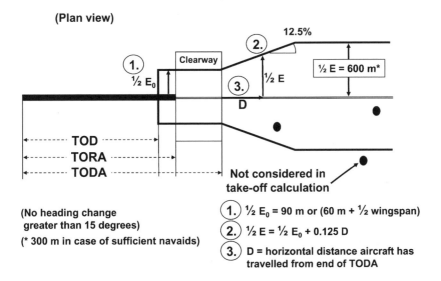

Figure 5.15 The Location of Obstacles JAR-OPS Definition

We now need to consider the final physical limitations, which are tyre speed and brake energy.

Tyre speed   Clearly, the highest rolling speed the tyres will reach on the runway is at the point of lift-off, at the speed V_{LOF}. This is the only speed specified in miles per hour, because tyres are produced for the car industry. High aircraft weight or V_2 results in high V_{LOF}. Once the tyre speed limit is attained, take-off speed and therefore permissible take-off weight would then be limited.

Brake energy   The maximum brake energy is the maximum speed from which a take-off can be aborted, called V_{MBE}. As weight increases, V_{MBE} decreases to ensure that the brakes can dissipate the energy expended in braking the aircraft. Once more, take-off weight might be dimensioned by this limitation.

Having covered the principal physical elements of the take-off, we shall now explore how ambient conditions play their role. This is where fleet planners have an opportunity to massage opinions of an aircraft's performance. Although we can play with assumptions in aircraft evaluation, of course there are strict rules that apply in real day-to-day operation.

Ambient Conditions

The efficiency of an aircraft and its engines is a function of the density of the air. The less dense the air, the less efficient the aircraft. Air density decreases as altitude and temperature increase, so at hot or high airfields an aircraft needs a longer distance to take off. Some of the classic 'problem' airfields are in Mexico, Madrid, Nairobi, Johannesburg, Addis Ababa, Harare, Sa'naa, Quito and La Paz.

It is obviously in our interests to correctly estimate the take-off weight at airfields where ambient conditions are challenging. However, the assumptions must be very carefully compiled in order that the results are not unrepresentative of real operating conditions. There are three areas where ambient conditions themselves can be moulded: temperature, elevation and statistical analysis. We shall now examine these, as well as some special operational practices that can also be applied to improve permissible take-off weight.

Temperature and elevation   Airfield take-off study assumptions should always specify the temperature at which the calculation should be made. The outside air temperature (OAT) is the actual ambient air temperature, which is frequently quoted in relation to a standard day, called 'ISA', or International Standard Atmosphere. At sea level, ISA is equivalent to 15°C. Thus, when the sea level outside air temperature is 30 degrees, then we refer to this as ISA+15. As elevation increases, temperature falls at a rate of 2°C per 1,000 feet, as a rule of thumb.

Problems occur when high elevation airfields happen to be fairly hot, such as in the examples quoted above. Manufacturers' generic take-off charts therefore reveal the range potential of an aircraft from high elevation airfields at high temperatures, this being the most critical case. Figure 5.16 illustrates a typical curve. Such generic data are particularly useful at the 'rough cut' stage of an analysis because it is easy

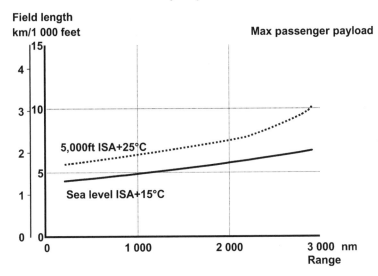

Figure 5.16 Field Length vs. Range – Example

to see whether, for example, the maximum passenger payload of the aircraft can be carried according to runway length, airfield elevation and temperature.

In this example, the aircraft would require a runway length of 7,500 feet in order to take off with sufficient take-off weight to fly its maximum passenger load for a distance of 2,000 nautical miles, if the airfield elevation were 5,000 feet and the take-off temperature were ISA+25, or else 30°C.

When quoting a temperature, we need to be clear if the value represents the hottest time of the day or not. There is a tendency for study assumptions to require that calculations be made at the temperature for the hottest time. When limitations occur, this should always be challenged, especially if it is certain that operations would take place at cooler times. Similarly, we must be aware if airfield temperatures are quoted for the hottest month or not.

Another trap concerns statistical probabilities. In an evaluation we cannot change airfield elevation, runway length or the position of the obstacles, but we can make a value judgement concerning the statistical reliability to be applied to the airfield temperature. For example, the average temperature over a period is referred to as 50% probability, meaning that on half of the occasions temperature will be higher, or lower, than the quoted value. Frequently, fleet planning assumptions require that 85% probability temperatures should be used. This means that on 85% of occasions the quoted temperature will not be exceeded. Thus, sufficient margin is applied to the temperature used for the take-off calculation to ensure that the resulting take-off weight can, in fact, be attained on the majority of occasions. Why 85%? Well, the real answer is lost in the folklore of fleet planning, although statisticians will note that it almost equates to the mean plus one standard deviation.

Another typical airfield temperature applied to performance studies is the 'average daily maximum', which is a measure given in the Aeronautical Information

Publications (AIP). The Chapter named 'AGA' (Aerodromes, Air Routes and Ground Aids) contains the data.

A fair way of dealing with the issue of airfield temperatures is to ask the manufacturer to make an assessment of permissible take-off performance by season or else by month. In this way, any significant variations in temperature can be identified.

Pressure altitude   Take-off capability depends a great deal on the pressure conditions, because engines deliver less thrust as pressure decreases, which is often the case for high elevation airfields. Although we were saying in the previous section that there is nothing that can be done to physical inputs such as airfield elevation; this is not strictly true. We must make a distinction between pressure and geometric altitude, which are identical only in standard atmospheric conditions. Pressure altitude on a given day is the altitude in ISA conditions corresponding to the real pressure measure on that day. Most of the time take-off calculations need to be considered in non-standard conditions. Confusingly, temperature variations can result in opposite variations in pressure for two different airports, making it impossible to conclude that there is a coherent relationship between temperature and pressure.

Clearly, when pressure altitude is below geometric altitude a higher take-off weight can be calculated, with a consequent improvement in permissible payload. It may sound like smoke and mirrors, but the use of pressure altitude is, in fact, the only correct way of assessing aircraft take-off performance. The formula for converting pressure into pressure altitude is shown below, valid below the tropopause (36,089 feet, or 11,000m altitude) and Figure 5.17 shows a worked example, based upon Harare airfield in Zimbabwe.

$$\text{Pressure altitude } (Z_p) \text{ in metres} = \left[1 - \left(\frac{P}{P_o} \right)^{\frac{1}{5.25588}} \right] \frac{288.15}{6.5 \times 10^{-3}}$$

Where P = the given pressure
Where P_o = 1013.25 hPa
(standard pressure measured in hecto-Pascals)

Thus, the application of pressure altitude has resulted in a 'gain' or reduction in altitude of up to 300 feet. This may improve the permissible take-off weight by several tonnes.

Wet vs. dry conditions   Standing water on a runway will obviously impede the progress of an aircraft across the surface. Energy that could have been used to accelerate the aircraft is dissipated and braking distances expand. The practical effect is that the required runway distances are increased. The regulations therefore ensure that friction coefficients are applied to the calculations and the screen height at the end of the runway is reduced from 35 feet to 15 feet. In addition, reverse thrust can be taken into account on wet runways. These last two corrections sometimes mean that the take-off performance calculation for a wet runway can actually be superior to that for a dry runway, which is legally forbidden. In this case the resulting take-off

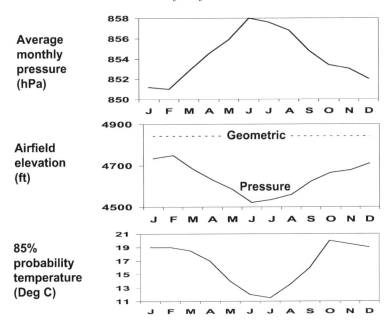

Figure 5.17 Pressure Altitudes at Harare

weight must be the lower of the two calculated values. Great care therefore needs to be exercised in interpreting permissible take-off weights.

Wind A headwind reduces the acceleration distance required because the aircraft can take off at a ground speed which is lower than the V_{LOF} speed. A tailwind has the opposite effect and the acceleration distance is increased. A pilot must take into consideration the effect of the wind component for a real take-off case. However, in aircraft evaluation it is common to consider still-air conditions. When wind conditions are to be calculated the usual method is to analyse the take-off in both directions on the runway and adopt the wind 'cross-over' point, which is the minimum take-off weight which could be expected, irrespective of the wind component of the two opposing directions of travel.

A tailwind is denoted by a negative sign. As the tailwind decreases and eventually switches to a headwind, so the permissible take-off weight increases. In most cases we need to consider a wind component of up to 10 knots. Figure 5.18 shows the crossover point calculation. You will see that there is a kink in the curve at the zero-wind condition. This is because margins are added for any wind calculation to take account of gusts. These margins are 150% of the calculated value for a tailwind and 50% of the calculated value for a headwind. The result is that a take-off weight improvement is less with a headwind than that with a tailwind for the same increment in wind speed.

A crossover calculation pre-supposes that the airline would be able to make a choice of take-off direction according to prevailing conditions. However, some airports may impose a preferred take-off direction to take account of local noise rules.

Take-Off Weight

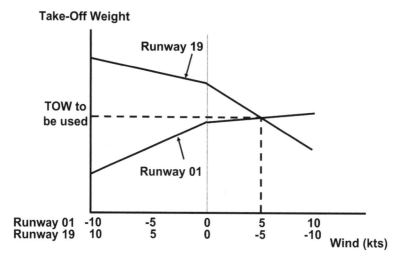

Figure 5.18   Take-Off Cross-Over Point

Operational practices to improve take-off performance   If, as a result of the application of statistical probabilities and pressure altitude, the take-off calculation is still insufficient, then we can consider the effect of other operational practices that may improve results.

Take-off weight can be significantly improved by using a turn procedure to avoid obstacles. The regulations require that an aircraft should not begin a turn until it has reached a minimum height of 50 feet. During a turn, performance is affected and the Flight Manual includes the necessary reductions according to the conditions. Sometimes, a turn procedure is required immediately after take-off where obstacles are too limiting. This is the case at Lhasa in Tibet, where the runway is located in a box canyon.

Another operating feature which increases permissible take-off weight would be to reduce the second segment climb-out gradient. Again, this can only be undertaken with the agreement of the airworthiness authority and under strict conditions. There should be an absence of obstacles in the take-off path, for example.

Runway Loading

A final ingredient for the fleet planner to be aware of is the load effect that an aircraft has on a runway. There are several methods used to assess runway loading but the most common is the ACN/PCN method, which has been adopted by ICAO and their member nations. A runway is assigned a Pavement Classification Number (PCN) and an aircraft has an Aircraft Classification Number (ACN). In order for an aircraft to be allowed to use a runway, its ACN should be less than the runway's PCN.

The ACN is a number that expresses the relative impact of an aircraft on a pavement of a specified standard sub-grade. A PCN is a number that describes the bearing strength of a pavement for unrestricted operations.

PCN is categorised according to pavement type (rigid or flexible), sub-grade strength (high, medium, low, or ultra-low), tyre pressure, and whether the evaluation method is based upon a dedicated technical appraisal or experience with the aircraft. The ACN is a function of the geometry of the landing gear (such as the distance between the wheels and the number of wheels), the weight of the aircraft, the tyre pressure, the centre of gravity, and the sub-grade category of the pavement. There are trade-off methodologies that enable movements to take place where an aircraft ACN exceeds the runway PCN. These 'overload operations' depend upon the percentage of overload movements of the total.

Although the ACN/PCN method is widely applied today, a number of other methodologies are in use. For example, the former ICAO method, called the LCN, or Load Classification Number, is still in use. There is even a method that is limited to a single aircraft type, developed by Boeing for the 777. Data for runways can be easily found in Aeronautical Information Publications. Aircraft data exists in the manufacturers' airport planning documents.

If an aircraft ACN exceeds the PCN at a number of airfields on a network, one solution could be to select a bogie option, if available. In return for a higher price, higher weight and some increase in maintenance costs, there may be opportunities to serve more markets. For example, the provision of a bogie as an option on the A319 enables that aircraft to land at 24 more runways in Australia and 28 more in Indonesia. This option can therefore enable that aircraft to be considered as a replacement of smaller regional jets serving less developed airfields.

It may well be the case that an apparent runway loading limitation, which reduces permissible take-off weight, turns out to be anything but a limitation. If, for example, the Maximum Design Take-Off Weight cannot be achieved, it is important to check that the required take-off weight in order to achieve all missions from that particular airfield is inferior to the runway limitation. If this is the case, then the runway-loading problem disappears.

We have now completed our examination of the main elements of an aircraft take-off which are of relevance to the fleet planner. The intent has been to highlight only those issues that directly concern aircraft comparisons. We shall now consider the *en-route* performance of the aircraft.

The *En-Route* Performance Analysis

To recap on where we are in the chain of analysis, we have performed a 'rough cut' choice of aircraft type based upon a market analysis, defined our aircraft configuration, assessed the operating weight of the aircraft, and established whether any take-off limitations might exist on the proposed network. Now we need to undertake a simulation of performance on the network in order to measure two principal parameters: the amount of payload that can be carried and the amount of fuel burned.

Like the take-off, aircraft *en-route* performance is very much a function of the aircraft definition and operational conditions. These we will examine in order to judge their influence on the result. One of the first problems we will encounter is that there are a number of interpretations of the expression 'aircraft performance'.

Nominal performance   This is the manufacturer's marketing level of performance and the one used in the performance analysis. It is the level of performance found in the manufacturer's Performance Manual, representing the average performance level expected at the aircraft delivery.

At first sight it may seem incongruous to use 'marketing' levels in order to assess real conditions. However, marketing levels do provide an honest and accurate estimate of how an aircraft is expected to perform. Importantly, marketing levels reflect future deliveries and therefore take into consideration any performance improvements that may be in the pipeline. Some improvements might come from airframe changes and some from engine fuel consumption improvements. In the latter case, the airframe manufacturer will audit any claims from the engine manufacturer before accepting the proposed changes in his own database.

It is not always the case that an airline will accept nominal performance at face value. If the airline believes that there is uncertainty in the ability of the manufacturer to deliver the levels of performance he claims, then there is always the possibility of adding mark-ups to the basic levels. We will categorise the mark-ups later in this Chapter.

Flight Manual and Flight Crew Operating Manual performance   A Flight Manual is produced for each aircraft and includes certified performance, such as the take-off and landing performance and various operating procedures. It is the official level of performance recognised by the certificating authorities and is the only level that can be used in real operational calculations, as opposed to fleet planning calculations. The Flight Crew Operating Manual (FCOM) is issued for each aircraft and engine model. It does not require the approval of the certificating authority and includes the certified performance found in the Flight Manual, plus non-certified performance, such as the *en-route* performance.

There are some problems which preclude the direct and systematic use of Flight Manual or FCOM performance in fleet planning. Firstly, the data do not include anticipated performance improvements, and secondly, the data are only available for aircraft that have been certificated.

Guaranteed performance   This category of performance does not strictly represent a real level of performance. Instead, it is a value proposed by the manufacturer to a client to take into account the risk of a certain level of performance not being achieved.

Performance guarantees are a contractual set of conditions between the manufacturer and customer. They are intended to protect the customer as well as limit financial exposure of the manufacturer. Performance analysis undertaken for a customer will often include fuel mark-ups and margins. However, these are provided in good faith and are not contractually binding. Guarantees, on the other hand, incorporate margins according to the degree of risk taken by the manufacturer, as well as penalty clauses for non-compliance.

Setting the Performance Analysis Parameters

There are many ways to skin a cat. Although the performance of an aircraft is a known quantity when ambient and operating conditions are defined, there are numerous ways of achieving the results sought for fleet planning purposes. Aircraft performance in fleet planning is as much a function of the assumptions and statistical probabilities as it is with the application of precise data. It is also dictated by available technology, sometimes not even connected to the aircraft, and by regulatory constraints.

For a performance analysis to have real value in a fleet plan, it is important that airline and manufacturer work in concert. It is in everyone's interest that agreement can be reached on how performance is calculated, presented and, crucially, used later on in a fleet plan.

The real objective of analysing performance   There are two principal objectives of calculating performance. Firstly, we are interested in knowing as precisely as possible the limits of what an aircraft can achieve. Having planned the configuration, estimated the aircraft weight and assessed the take-off performance, we need to know what the limits of payload and range actually are. The payload-range curves will not help us understand the true limits of an aircraft's performance on a real route because they do not consider the effect of wind and are based on a unique alternate airfield reserve policy. Yet it would be a mistake to base all future decisions on what an aircraft can achieve at the limit, because that would artificially colour our impression of the aircraft's ability to perform in less onerous conditions. So, the second objective is to make an assessment in an operating environment that is more representative of average conditions.

The first objective is satisfied by setting ambient conditions at the 'critical' level, and the second objective is achieved by resetting the conditions to 'average' levels. Hence, we will need to do all our performance analyses twice. In addition to changing the ambient conditions, we shall also need to revise the load factor of the aircraft. For the critical performance case, we should consider a full aircraft, carrying the maximum number of passengers and maximum amount of cargo according to the airfield conditions. For the average performance case, we must assume an average load factor.

To summarise, for critical performance we would typically assume 85% probabilities for airfield temperature and *en-route* wind. This is to say that the values used would be valid for 85% of occasions. The critical case should assume maximum structural or volumetric payload, according to which is most limiting. Average performance should take into account 50% airfield temperatures and *en-route* winds, and average expected payloads for both passenger and cargo. Typically for a scheduled airline, a passenger payload of 75% and cargo payload of 50% would be applied. It is the fuel burn and payloads associated with the average case that form inputs to the economic study.

Tracks   One would imagine that the shortest distance between two points is a straight line, or the great circle. Regrettably, fleet planners must deal with a wealth of often conflicting data sets which suggest that getting from A to B is anything but straightforward.

Aircraft mostly follow designated airways, Standard Instrument Departures (SIDs) and Standard Terminal Arrival Routes (STARs). Where no designated airway exists, random tracking can be used. In some areas of the world even the designated tracks actually shift position twice a day. Clearly, the complexities and highly variable nature of

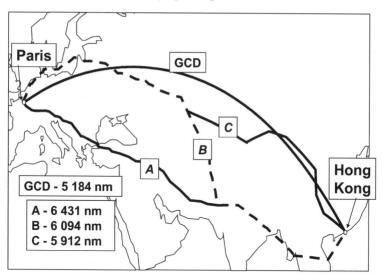

Figure 5.19   Track vs. Great Circle Distance

real aircraft operations are very difficult to replicate in a single performance calculation, which is intended to be used as the template for airline payload capability.

There is nothing to stop the manufacturer applying for a flight plan in order to ascertain a real airway routeing, and use this as the basis for the performance calculation. However, where wide variations in available routeings occur, it is important that the airline advises the manufacturer which airway it habitually uses. As an example of the dramatic differences, consider the sample of airway tracks available to get from Paris to Hong Kong, shown in Figure 5.19. The ratio between the longest track in this example and the great circle distance is 24%. In fact, there are approximately 20 different tracks to choose from. Real operational conditions that can result in a variety of tracks can range from traffic density, wars and other political considerations, or even a lack of overflight rights. Pinning down aircraft performance in fleet planning is not an exact science.

The way around this problem is to apply the irrefutable logic of taking the great circle distance, but with an agreed percentage addition to take account of different airway distances and circuitous take-off and arrival patterns. This method has the added advantage of ensuring that any performance comparison made between data emanating from different manufacturers is based upon similar distances. The main disadvantage is that we have already given up any attempt at replicating a real operation.

If we intend to apply mark-ups to great circle distances, then there is some merit in applying different mark-ups for short-haul compared to long-haul studies. If the short-haul system warrants, say, a 5% mark-up on great circle distance, perhaps the long-haul operation might be unduly penalised, especially if the actual airways are fairly close to the great circle.

The organised tracks   There are two principal zones of the globe where organised track procedures apply: the North Atlantic and North Pacific. The systems are called

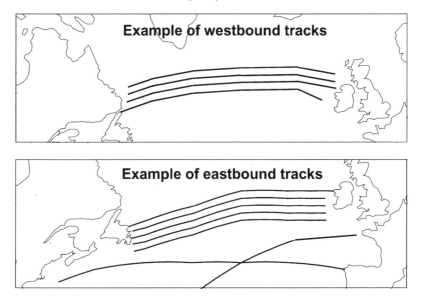

Figure 5.20 The Organised Tracking System

Organised Tracking System (OTS) for the former and Pacific Organised Tracking System (PACOTS) for the latter.

The essence of the OTS is that both traffic and meteorological conditions are very specific. Owing to time zone differences traffic tends to leave Europe in the morning in order to arrive in North America in the afternoon (local time). Traffic from North America leaves in the evening in order to arrive in the European morning. There are thus significant flows of uni-directional air traffic in both directions for defined periods. The organised tracks are assigned twice daily, with Gander Oceanic Control responsible for the eastbound (night-time) OTS and Prestwick responsible for the westbound (day-time) system. The published tracks vary according to the position of the jetstream and comprise sets of coordinates, plus defined entry and exit points on either side of the Atlantic. Lateral separation of tracks on the North Atlantic is 60 nautical miles. It is not obligatory for aircraft to follow the OTS. Aircraft can use random tracks if they are flying outside of the tracking system, or if they are flying at times when the system is not in operation.

Understanding the operational practices on the North Atlantic is important for fleet planners because fuel calculations between city-pairs may vary due to the constant movement of the tracks. It is therefore important to establish whether fuel calculations should be based upon 'preferred tracks', taking into account the most usual routeings available to an airline. The position of the most frequently used track will also have an impact on the wind strength used in the fuel planning. Figure 5.20 shows an example of the OTS.

Vertical separation This had historically been 2,000 feet between flight levels, but in an effort to increase capacity Reduced Vertical Separation Minima (RVSM) was progressively introduced which reduces the separation to only 1,000 feet. RVSM

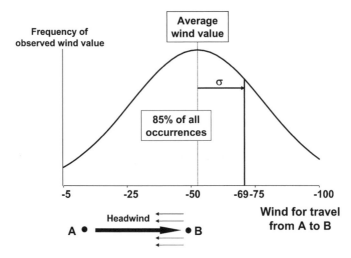

Figure 5.21 Wind Probability – Headwind

is applicable between Flight Levels 290 and 410 (29,000 feet and 41,000 feet) and has now become applied extensively throughout the world. The benefits of RVSM are reduced fuel burn owing to opportunities to better optimise flight levels and also reduced delays as more levels are available to controllers. China and Russia use metric flight levels requiring adjustments to be made to flight levels.

✈ *Temperatures and winds* We have already examined the effect of temperature and wind on the take-off calculation. These two elements also affect the *en-route* performance. Temperature influences engine performance and therefore the rate of climb of the aircraft. It also affects cruise performance where there is a deviation from ISA (standard) conditions. Wherever temperature deviates from ISA, geometric and pressure altitudes differ as well. Temperatures higher than ISA result in the geometric altitude exceeding the pressure altitude. Thus, an 'ISA+10°C' temperature at 33,000 feet pressure altitude is equivalent to 34,300 feet geometric altitude. This means that, compared to standard conditions, more energy must be expended in order to lift the aircraft to the higher altitude.

✈ Wind has a more dramatic effect on aircraft *en-route* performance. If the wind component does not change with altitude, then the time taken and fuel burned to the top of the climb will remain the same. In practice, the wind component often does change with altitude, so the climb angle and flight-path angle do alter, changing the distance travelled over the ground. However, of greater significance is the influence of wind on the cruise. In the case of a headwind, the air distance becomes greater than the ground distance and more fuel is consumed, and in the case of a tailwind the opposite is true. For each performance calculation we need to determine the Equivalent Still-Air Distance (ESAD) for the mission. The ESAD is equivalent to the ground distance only in zero-wind conditions.

The elements that determine the ESAD are the amplitude of the wind, the altitude and statistical probability. The first two elements are relatively easy to determine.

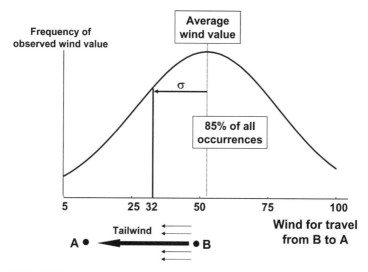

Figure 5.22 Wind Probability – Tailwind

It would be quite usual to apply seasonal winds, such as summer or winter, or else month by month. Altitude has a lesser effect, but there are some differences between the most commonly used flight levels.

Statistical probabilities are applied to wind amplitude in the same way that they are to airfield temperatures. As an example, let's take an 85% never-exceed wind probability that might result in a 69-knot headwind in one direction and a 32-knot tailwind in the return direction. We presume that the statistical distribution of wind follows the normal, or Gaussian curve, so our 85% probability virtually equates to one standard deviation. The effect is shown in Figures 5.21 and 5.22.

Two values of wind probability are typically applied. As with airfield temperature, we need to know the 'critical' level of wind (usually 85% probability) so as to be sure that on the majority of occasions the payload can be carried on a route. However, this rather high 'never-exceed' level is not at all reasonable for the determination of payloads and fuel burns for economic calculations, so we need to calculate the 50% level as well.

The Flight Profile

We have seen that, despite detailed knowledge of how an aircraft should perform in take-off and during a mission, aircraft performance is turning into something of an art rather than a science when it comes to fleet planning. Much depends upon the ambient conditions and the degree to which statistical probabilities are applied. Determining the flight profile is no exception to this trend.

A real aircraft dispatch in a flight plan is not directly related to the flight profile in the context of fleet planning. The objective of the flight profile is to group together the most typical operating conditions so that payload and fuel burns and flight times can be assessed. Having said this, flight profiles must be agreed with the local airworthiness authority, which may decide to apply standard profiles produced by either the FAA or the JAA. The local authority may even decide to exceed the FAA or JAA levels. In the final analysis, it is the Captain who decides on the fuel to be loaded, but the airline fuel policy can help assess additional fuel loads for specific cases.

Flight and block times   The first distinction we should make is between flight time and block time. The block time, as the word suggests, describes the time taken between 'blocks off' and 'blocks on' – in other words, engine start-up, ground manœuvres, or taxi times, are included. The flight time incorporates the mission time from start of take-off run up to the landing of the aircraft. The distinction is important as one economic calculation, maintenance cost, is based upon flight times whereas the others are based on block times.

Climb, cruise and descent   Every flight profile incorporates a take-off, climb, cruise, descent and landing. Beyond this there is a requirement to allow for a climb, cruise and descent to a nominated alternate airfield. However, there are numerous permutations possible. Many climb profiles are not continuous as, according to the weight of an aircraft, the optimum performance can be better achieved through a step climb, which means that there is an intermediate cruise at a lower than final altitude whilst fuel is burned off to optimise the last part of the climb.

It is usually presumed that the climb portion starts at a pressure altitude of 1,500 feet above airfield elevation and, similarly, the descent ends at 1,500 feet above the elevation of the arrival airfield. No credit for take-off and approach distances is taken in *en-route* calculations. Thus, the mission distance starts and ends at the 1,500 feet elevation points.

The flight profile should also specify the flight levels to be applied and the aircraft speeds in climb, cruise and descent. Most aircraft can climb and descend in a variety of configurations, depending upon whether time or fuel is being optimised. Cruise altitudes are usually subject to semi-circular rules. These indicate the normal altitudes according to whether the aircraft is moving in a westerly or easterly direction. At cruise altitudes, the distance between flight levels for one direction of travel is nominally 4,000 feet, although this has largely being supplanted by the introduction of Reduced Vertical Separation Minima allow the halving of separations between Flight Levels 290 and 410. Unless RVSM applies, in aircraft performance analysis it is common to apply the 'semi-circular' rules, where only odd flight levels are assigned depending upon the direction of flight. Thus, for track headings of 000–179° flight levels FL290, 330, 370, etc. are used, and for track headings of 180–359° flight levels FL310, 350, 390, etc. are used. Countries with predominantly north to south traffic, such as France and New Zealand, have semi-circular rules defined in a north:south fashion as opposed to an east:west fashion.

Reserve Policies

Airworthiness authorities require an agreed reserve policy that should incorporate an allowance, fuel to reach a diversion airfield, and a holding policy. The purpose of a reserve policy is, naturally, to ensure that the aircraft is carrying enough fuel to reach another airfield in the event of the destination airfield being unavailable. Reserve fuel also includes various margins on the basic calculations for the mission fuel.

Selection of alternates Flight planning rules require that each flight include a nominated alternate airfield in case it is impossible to land at the intended destination. Diverting to an alternate airfield usually occurs owing to poor weather conditions, but may also be due to an obstruction on the runway. Thus, the fuel taken on board at the beginning of a mission will be dimensioned by the chosen alternate airfield. The longer the distance from the intended destination to the alternate, the higher becomes the amount of fuel to be loaded at departure. Also, in our discussion on the payload-range envelope (Figure 5.7) we saw how the Maximum Landing Weight of the aircraft might become payload limiting if the reserve fuel policy becomes very stringent. This is because the amount of fuel required to ensure that a long diversion can be attained grows beyond the MLW limit of the aircraft.

Thus, the choice of alternate airfield becomes significant in a performance calculation. Selection of an alternate that is too close to the intended destination can be considered inappropriate because it can be argued that the weather conditions at both airfields could be similar. On the other hand, selecting a very distant alternate may invoke the MLW limitation of the aircraft. In any case, a long alternate means that fuel burn is increased, as the aircraft must burn fuel to carry fuel.

Much depends upon the geographical region, the stability of weather patterns and the presence of airfields. A good rule of thumb is to select an alternate airfield that is between 150nm and 250nm of the destination. A payload-range calculation is based upon a single choice, but a real route must always include a nominated diversion.

Allowance The reserve fuel includes an allowance, or contingency fuel, to take account of factors that were unforeseen in the planning stage of a mission. These factors include: deviation of an individual aircraft from the expected performance level; *en-route* winds which may have been higher than expected; and deviations from planned routeings or flight levels. The allowance is often based upon a percentage of trip fuel for the mission, or the mission flight time.

As an example, the JAR-OPS Fuel Policy (AMC OPS 1.255) requires that the contingency fuel be equivalent to 5% of the planned trip fuel. This can be reduced to 3% if an *en-route* alternate is available within a radius of 20% of the total flight plan distance. The centre of this circle must be situated at a point along the planned route that is either 25% of the mission length from the destination, or else 20% of the mission length plus 50nm, whichever is the greater. Figure 5.23 shows the requirement.

In addition to the above opportunities for reducing allowances, there are two additional mechanisms permitted under JAR-OPS. Either the allowance can be equivalent to 20 minutes' flying time, or else 15 minutes holding above destination, so long as the operator has an established fuel consumption monitoring programme backed up with statistical analysis. It sounds complex, but it is worth persevering as it can often result in a reduction in allowances carried. The technique is thus useful for both aircraft operation as well as evaluation.

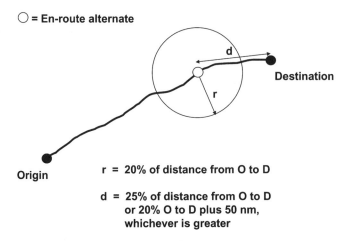

○ = **En-route alternate**

Destination

Origin

r = 20% of distance from O to D

d = 25% of distance from O to D
or 20% O to D plus 50 nm,
whichever is greater

Figure 5.23 JAR-OPS Fuel Policy

The diversion profile   A missed approach, or overshoot of the destination airfield, is usually calculated based upon 80% amplitude of a normal take-off at the weight of the aircraft at that moment. Then, the aircraft is presumed to climb to a somewhat lower altitude than that of the mission. A diversion altitude of either 20,000 feet or 25,000 feet is usual, and the distance to the diversion airfield may or may not take into consideration winds.

Holding   After a descent towards the alternate airfield there should be provision for a holding. This usually amounts to either 30 or 45 minutes and may be presumed to be accomplished in straight and level flight or, more accurately, adopting a 'racetrack' pattern to take turns into account. Aircraft performance in a holding pattern also differs according to the configuration. A 'minimum drag' configuration is preferred in order to minimise fuel burn.

Island reserves   Aircraft flying to island destinations may be faced with the prospect of nowhere to go in the event of a diversion. The principle purpose of the diversion fuel is to enable the aircraft to land safely in the event of a change in the weather or else a runway being blocked at the intended destination, but an island destination carries a different level of risk. Accurate weather forecasts and monitoring should ensure that an aircraft never arrives over its island destination to find that landing is impossible due to bad weather. In the absence of a nominated airfield, the provision of two hours' fuel for continued cruise is usually permitted to ensure that any impediment to landing can be removed. Figure 5.24 shows a typical profile.

Summary of Definitions

Segment or stage fuel	The fuel burnt during climb, cruise and descent.
Trip fuel	Fuel burnt during take-off, climb, cruise, descent and approach.
Block fuel	Fuel burnt during the trip, plus taxi-out and taxi-in.
Total fuel loaded	Fuel burnt plus allowance plus diversion.

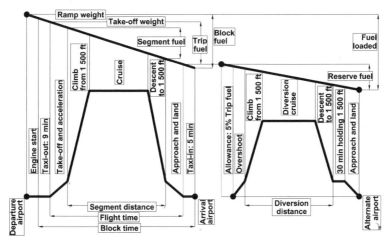

Figure 5.24   A Typical Flight Profile

The same definitions apply for distance covered and time.

Special Practices

The majority of route performance calculations can be based upon the parameters described so far in this Chapter. However, there are two special practices that may be applied on some types of operation that might be limited: tankering and reclearance.

Tankering   It is usual to presume that an aircraft will be fuelled before each sector it flies. However, there are some airfields where either fuel supplies are unreliable or there is no fuel supply available at all. For example, Lhasa airfield in Tibet does not have any fuel supply, so every incoming flight must carry sufficient fuel for the next stage plus, of course, sufficient reserves appropriate for that sector. Apart from cases of obvious necessity, there are also economic reasons why an operator may elect to 'tanker' fuel. The economic reasons are two-fold. Either an operator may seek rapid turn-rounds, which is particularly important for multi-leg flights in the commuter market, or else the fuel price at the destination is so high that it makes economic sense to curtail as far as possible the uplift of fuel.

Carrying fuel for an onward stage obviously imposes some penalties. The first of these concerns the structural limits of the aircraft. We have already seen that the Maximum Landing Weight is one of the key parameters in determining aircraft performance (Figure 5.7). The MLW also dimensions the amount of extra fuel that can be tankered, as seen in Figure 5.25.

In this case, a Maximum Take-Off Weight limitation further limits the fuel that can be tankered for a given payload. When carrying a large payload, a tankering procedure is likely to be limited by MTOW or MLW. On long range operations with lower payloads the limitation is more likely to be the Maximum Fuel Capacity of the aircraft.

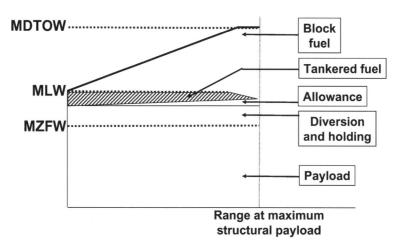

Figure 5.25   Tankering and Range

The second penalty which occurs is that payload can be reduced. For a given range there is a trade-off between the amount of fuel which can be tankered and the payload carried. Figure 5.26 describes this relationship, which is expressed as a percentage of structural payload versus round trip fuel. There is an economic balance to be struck between the price difference of fuel between the origin and destination, the value of payload surrendered, and the cost of fuel burned to carry additional fuel.

Sometimes, an operator is forced into tankering for bizarre reasons. During the blockade of fuel supplies by protesters in France in 2000 some airlines tankered fuel to French destinations in order to be able to exit the country. The resulting payload limitation on some flights meant that the passengers, but not their luggage, were carried!

Reclearance   We have seen that part of the allowance includes a percentage of trip fuel, resulting in high allowances for long missions. However, improved weather forecasting accuracy means that such high reserves are rarely needed. Reclearance is an operational procedure designed to decrease the amount of reserve fuel loaded for a mission.

Figure 5.27 outlines the concept of reclearance. The upper diagram is a simplified flight profile of the intended flight plan. The fuel carried under this 'normal' reserve policy would consist of the fuel required from Airport A to Airport B, plus the trip fuel allowance and the diversion fuel to the alternate airfield. However, a reclearance procedure involves a flight plan that considers a reclearance point, D, and an *en-route* alternate, Airport C, as the phantom destination. The alternate for Airport C, C_1, may or may not be Airport B.

The fuel carried in a reclearance operation should be when the two cases in Figure 5.27 are equivalent. If the allowance fuel up to Airport C (presumed in the

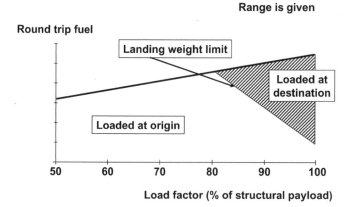

Figure 5.26 Tankering and Payload

example to be 5%) has not been burned at the reclearance point, then the flight can continue to the final and intended destination.

There is an optimum reclearance point for a mission, depending upon the distance to the alternate and whether the initial destination is along the track. A typical optimum reclearance point would be around at 90% of the trip distance, depending upon the allowance. Reclearance is often used when fuel tank capacity is insufficient for the normal mission and reserves, or when the intended payload cannot be accommodated due to the fuel requirement.

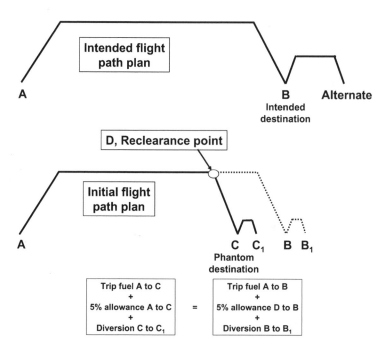

Figure 5.27 Reclearance Flight Profile

ough reclearance is an operational procedure, it can be used in performance
ions in fleet planning where a normal flight profile would otherwise paint an
inaccurate picture of what an aircraft could actually achieve.

Extended Twin Operations

There are special regulations that govern the design and operation of twin-engine
aircraft to enable them to fly further than one hour from an airfield at their one-engine
inoperative speeds. Extended Twin Operations, or ETOPS, is now a commonplace
and successful practice and the track record of success for both manufacturers and
operators is excellent. However, there are significant differences in the way in which
twins and multi-engine aircraft are operated, so the issue is highly relevant in fleet
planning today.

The evolution of ETOPS   A huge effort took place in the 1980s on the part of the
manufacturers, airworthiness authorities and operators to put the ETOPS regulations
into place. Until that time, twins were limited to routes that took them no more than
60 minutes from any airfield, at their one-engine inoperative speed. This regulation
dates back to 1953 and, whilst relevant to the piston-engine era, had become an
anachronism by the 1980s. Indeed, there was a mismatch between the regulations
in force and the rapidly expanding payload and range capability of twin-engine
aircraft. Airlines were able to partially exploit their twins by applying the ICAO
recommendation of flying up to 90 minutes from an airfield at the all-engines operating
speed. This is equivalent to around 105 minutes at the single-engine speed.

In 1985, the FAA Advisory Circular 120-42 permitted twins to fly up to 120
minutes from an airfield, extended to 180 minutes in 1988. In the space of a few
years, vast numbers of city-pairs were opened up to twins, especially over the North
Atlantic, the Tasman Sea, and over many land masses as well.

This sudden explosion of opportunity did not come for free. Firstly, the aircraft
themselves had to be deemed ETOPS-worthy. Airworthiness authorities issue an
ETOPS Type Certificate for each eligible airframe-engine combination. Existing
designs, such as those of the 767 and A310 models, not only needed some modification
but also had to be proven in non-ETOPS service before receiving their ETOPS
approvals. Initial ETOPS aircraft had to accumulate up to 250,000 engine flying
hours, or sometimes even more, before they were deemed to be ready to fly ETOPS.
Derivative aircraft and engine designs were able to take advantage of commonality
in order to significantly reduce this requirement, sometimes down to zero hours,
depending upon the degree of technical similarity and experience.

Reliability and ETOPS   Underpinning all of the above is reliability. Each airframe
and engine combination of aircraft eligible for ETOPS had to be assessed in terms
of engine in-flight shutdown rates. The key to ETOPS is that any failure of a second
engine on a twin is presumed to be independent to that of the first failure. Thus, a
flameout due to an aircraft running out of fuel, for example, is not considered as an
ETOPS-specific event as the number of engines would be immaterial.

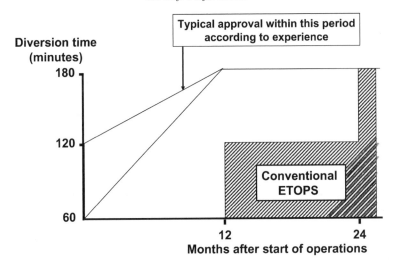

Figure 5.28 ETOPS Accelerated Approval

As ETOPS experience has continued to be amassed, the reasons for aircraft diversions became more associated with the systems on the aircraft rather than with the engines themselves. ETOPS aircraft must have a level of systems redundancy equivalent to that of multi-engine aircraft.

Second-generation ETOPS aircraft, the 777 and A330 families, were designed and built with ETOPS in mind, with no specific modifications being required for them to be considered for ETOPS operations, although an ETOPS Type Certificate is always subject to a continuing review of reliability. In 2001, the FAA allowed a 15% extension to the 180-minute diversion limit to enable the 777 to fly North Pacific routes in the event of diversion airfields being unavailable. The exemption is limited to the 777 and is subject to specific operating practices.

Operating under ETOPS   The original FAA Advisory Circular, echoed by a number of other major regulations on ETOPS issued by the British, French, Canadian, Australian authorities and, latterly, the JAA, contains the stringent operating procedures required for ETOPS.

It has never been possible for the operator of a twin to simply acquire an ETOPS-certified aircraft and then, with no previous experience of ETOPS, start flying as such immediately. The granting of an Operational Approval is contingent upon certain criteria being met. For example, the operator must have knowledge of the zone of operations, the aircraft type itself, dispatch procedures under ETOPS, an ETOPS-specific maintenance programme, a system for monitoring ETOPS flights, as well as adequate training procedures. The basic rule requires that inexperienced operators accumulate 12 months of non-ETOPS experience before being granted a 120-minute approval, and then fly another 12 months before moving to 180 minutes. This waiting period of up to two years before ETOPS can be applied is not acceptable to most new operators of ETOPS aircraft. The authorities can therefore grant an earlier Operational Approval, illustrated in Figure 5.28. Such a programme may embrace some form of simulated ETOPS dispatch on non-ETOPS flights, plus accelerated training. Another mechanism to get going quickly would be for a new operator to contract-out the ETOPS operation to an experienced airline for the initial period of operation.

An ETOPS operator must consider a number of elements. These include the status of *en-route* alternates, the area of operations for ETOPS, dispatch weather minima, the Minimum Equipment List (MEL), the various diversion strategies applicable to ETOPS, and critical fuel requirements. All of these elements impinge upon the twin, rather than the multi-engine aircraft, and all have varying implications for the fleet planner in assessing operational risk and economics. We shall now consider each in turn.

Status of en-route *alternates*   For the purposes of ETOPS, airfields are divided into two categories: adequate and suitable. An adequate airfield is one that possesses sufficient technical support, such as navigational aids and fire-fighting capability, and can accept the aircraft from the performance point of view. A suitable airfield, on the other hand, is one which is deemed to be adequate, but at which the weather minima for landing are at or above the levels required for ETOPS for a period of one hour before and one hour after the estimated time of arrival of a diverting aircraft.

There is no such thing as a definitive list of adequate airfields. Each national airworthiness authority can exercise its own judgement as to whether a particular airfield is adequate for operators under its jurisdiction. Indeed, individual authorities have frequently disagreed as to whether particular alternates qualify as 'adequate' or not. For example, Narssarssuaq is situated at the southern tip of Greenland and is well positioned as an *en-route* alternate for North Atlantic ETOPS operations. However, most regulatory authorities consider that the difficult approach to the airfield, coupled with poor weather and restricted navigation facilities, exclude it from any list of adequate airfields. A similar debate rages over the operational suitability of diversion airfields in the North Pacific, many of which suffer from extreme weather conditions.

Area of operations   An aircraft track under ETOPS must stay within the maximum authorised diversion time permitted by the national authority for both the aircraft type and operator. The area is defined according to isochrones, or time circles, based upon a set of adequate airfields.

The radius of these circles takes into account an individual aircraft's single-engine performance. The calculation is made based upon the estimated gross weight of the aircraft at various critical points within the area of operations, together with the optimum altitude. Credit is taken for drift down procedures. The aircraft Flight Crew Operating Manual provides the necessary data for the operator to determine the isochrones, and therefore the area of operations.

An aircraft is deemed to have entered into the area of operations once it has crossed the ETOPS entry point (EEP), where it is further than one hour's flying time from an airfield at the single-engine speed.

Figure 5.29 shows an area of operations for an ETOPS operation having a maximum authorised diversion time of 180 minutes. At no time can the aircraft stray into the unauthorised zone, of course.

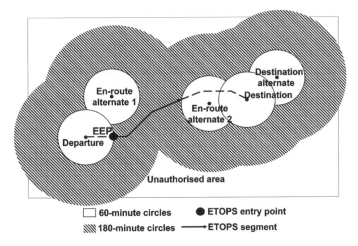

Figure 5.29 ETOPS Area of Operations

Weather minima for landing From the very beginning of ETOPS it was accepted that conventional landing minima should not apply for diversions to alternate airfields. Therefore, a more conservative approach is taken for the purposes of ETOPS flight planning.

Minimum Equipment List (MEL) An operator's MEL is based upon the Master Minimum Equipment List (MMEL) which is approved by the relevant airworthiness authority. An MMEL for ETOPS includes certain additional items according to the maximum approved diversion time. These items are many and varied, and depend upon a safety analysis of crew workload and the availability of critical systems. For example, for certain diversion strategies and certain aircraft types, the Auxiliary Power Unit must be operational at dispatch.

ETOPS diversion strategies This is an important area where there is a clear difference between the operation of a twin and a multi-engine aircraft. ETOPS regulations state that, in the event of an engine failure, a twin must divert to the nearest suitable airfield. However, the Captain of a multi-engine aircraft has the option of continuing the flight.

Performing a diversion to the nearest suitable airfield may have considerable commercial implications if the airfield is ill-equipped. Passengers need to be ferried out and the aircraft has to be recovered too, with significant economic and commercial ramifications.

Operators can select their own speed and fuel strategies for performing a diversion, according to their own particular needs. Designing a diversion strategy based upon high speeds means that the diversion circles of adjacent airfields overlap more easily, giving more flexibility, although more fuel would be consumed.

ETOPS critical fuel Here is another area where a twin carries a penalty compared to a multi-engine operation. The regulations require that enough fuel be on board the

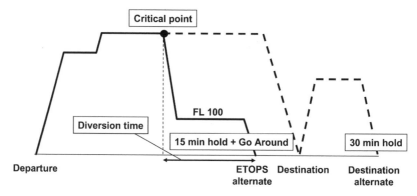

Critical point is where a simultaneous engine
failure and decompression are assumed to occur.

A comparison must be made between the standard fuel planning and
the ETOPS critical case. Actual fuel loaded is whichever is the greater.

ETOPS alternate and destination may be the same.

Figure 5.30 ETOPS Critical Fuel

aircraft to enable a diversion from the most critical point of the mission, presuming a simultaneous engine failure and decompression. This scenario could occur if an engine blade penetrates both the engine casing and the fuselage. In this case, the aircraft would need to descend rapidly to Flight Level 100 (10,000 feet) and fly at that level, on one engine, to an airfield which may be located anywhere up to the maximum diversion time authorised.

A twin-engine aircraft flying on one engine at 10,000 feet burns fuel at a faster rate than flying on both engines at normal cruise altitude. In addition to this penalty, the regulations require significant margins to be added to the fuel calculation for such a diversion. Depending upon the specific case these margins include: 5% contingency; a mileage penalty of 5% (or demonstrated performance factor); the adverse effect of any deviation from the MEL; anti-ice systems running; an allowance for ice-accretion on unprotected surfaces; 5% weather avoidance (for diversion times greater than 138 minutes); and APU fuel consumption, if the APU is considered as a power source. This remarkable collection of mark-ups can total over 20% of the nominal calculation.

The result is that the standard fuel planning may not be sufficient to cater for the so-called ETOPS 'critical scenario'. In this case, an additional amount of fuel should be added to the fuel loaded to ensure that the critical scenario can be met. Typical values are between one and five tonnes of additional fuel to be loaded, depending on the diversion time being considered and the distance from the destination to the destination alternate. In view of the magnitude of the penalties, many in the industry support initiatives to reduce the mark-ups. This can be achieved by reducing ice-related penalties when no ice is forecast, for example.

Figure 5.30 shows a simplified flight profile with the ETOPS critical fuel effect.

The economics of ETOPS   When ETOPS first burst upon the scene in the mid-1980s many operators found that they were experiencing an increase in costs. This was hardly surprising as the concept was new and investment had to be made in training, maintenance and reporting systems, and ETOPS-specific spares. Such has been the success of ETOPS that these investments have paid off in terms of better reliability. Also, ETOPS opened up new opportunities for airlines that would have otherwise remained closed to this category of aircraft.

With the coming of the second generation of ETOPS aircraft in the 1990s the concept had become so established that it was becoming increasingly difficult to identify costs specific to ETOPS. However, there are some cost areas that are very clear. We have seen how critical fuel scenarios require more fuel to be loaded, which can translate directly into less payload for routes which are limited by take-off weight.

Another significant area of cost concerns diversions. Depending upon the location and logistical challenges an ETOPS diversion can entail costs of over $1 million. Of course, it can be argued that some diversions, such as medical emergencies, would take place irrespective of the number of engines on the airframe. Nevertheless, so long as there are rules that govern specific conditions under which a twin *must* divert, an ETOPS operation carries a different level of risk.

The future of ETOPS   There is no question that ETOPS has become a commonplace issue, bearing testimony to the extraordinary efforts made by the manufacturing and regulatory community who, together with operators, have ensured an exemplary safety record. Indeed, the fact that both Boeing and Airbus have designed their third-generation mid-market aircraft families around twin-engined aircraft is proof enough that the concept is here to stay. However, debate continues to rage as to what the upper limit of ETOPS might actually be. In Europe there has been an initiative to introduce LROPS (Long-Range Operations) as a way of embracing all aircraft engaged in long-range operations, irrespective of the number of engines. LROPS is based on the principle that the flight crew should be able to select the safest alternate airfield to land and not necessarily the closest. LROPS-equipped aircraft would be equipped with enhanced medical equipment, fire-fighting equipment and communications systems.

The United States viewpoint differs. Various bodies have promoted the extension of ETOPS beyond the current 207-minute limit to 240, or even 330 minutes. This would enable not only North Pole, but also Antarctic and South Pacific operations as well. However, the JAA (now EASA) did not agree with the proposals submitted by the FAA for public comment in 2004 and a stalemate ensued. Consequently, EASA has produced its own draft for flights beyond 180-minute diversion times.

It can be argued that regulations based purely on time are outmoded. The industry would benefit from a regulation which addresses all aircraft, not just twins, and that considers the variation in operating conditions in different areas of the world. A new classification of airfields would be more appropriate, so that safety after landing, for example, is considered.

There is no doubt that ETOPS has been a tremendous success for the industry and that safety in all areas has been enhanced due to the increased scrutiny required of aircraft design, reliability and operation. ETOPS must form part of a fleet planning strategy, as significant differences still exist between the requirements of operating twins and multi-engine aircraft, and this has an effect on the economics of operation.

Combining Probabilities

To complete this Chapter we shall address a simplified approach to weighing the effect of the multitude of parameters under study.

When studying a large variety of variables it is often difficult to judge their combined effect on the result. For example, a route performance analysis result may be dimensioned by both the airfield temperature, which may limit the take-off weight and therefore payload, as well as the *en-route* winds, which may further limit (or improve) the payload as the equivalent still-air distance lengthens (or shortens).

Another common combination of variables is that of passenger demand and *en-route* wind. In this case, we need to know whether the predicted load matches the passenger demand, and how much residual capacity is available for cargo.

One way of assessing the combined effects is to study them in sequence and record their impact. Alternatively, the overall effect could be lessened by changing the individual probabilities of airfield temperature and *en-route* wind. However, neither of these approaches can take account of the fact that the variations act in concert, even though they may be independently driven.

A common analytical practice, therefore, is to combine the statistical probabilities by inducing a Monte Carlo simulation. This involves simulating sets of variables by a series of trials, or events. The greater the number of trials, the more likely we can achieve a result closer to reality. It would be typical to use around 1,500 trials for statistical security.

The Monte Carlo simulation generates random values of (say) wind magnitude and passenger demand, based on a mean value and standard deviation. The combination of the results can enable us to establish more realistic actual payloads, cargo potential, whether technical stops are necessary, and to identify on how many occasions passengers might be refused.

Naturally, the usual 85% annual wind probability and maximum payload capability are still important, particularly when comparing aircraft types. However, a much closer appreciation of real aircraft performance can be obtained by a Monte Carlo simulation.

Does the Aircraft Really Fit?

It is useful to perform a number of sensitivities of the performance analysis. Changes in Mach Number, seating configuration, operating weights, flight levels, and routeings can all affect the final results.

At some stage the results need to be compared to the real need of the operation. If the aircraft over-performs on the majority of the routes tested, then questions may need to be raised about the take-off weight option studied, for example. Demand in scheduled markets can vary dramatically as a function of time, both by time of day and between seasons. The aircraft performance should capture as great a proportion of the forecast market as possible without generating too much excess capacity. A great deal of intuition needs to be applied to ensure that all the inputs to the performance analysis give results which are meaningful for both aircraft comparisons and apply to a real operation.

In Summary

In this Chapter we have reviewed aircraft performance analysis to a level of detail appropriate to the fleet planner. Our goal is to correctly build-up the weight of competing aircraft types for study purposes, so that structural and volumetric payloads can be assessed.

Then, we examined the forces that shape the payload-range envelope. This all-important diagram has little application in day-to-day aircraft operations, but is of considerable use in comparing aircraft capability in standard conditions.

Analysis of aircraft performance is best divided into airfield and *en-route* performance. The correct assessment of take-off performance requires an appreciation of regulations as well as the physical and ambient limitations of specific airfields. The permissible take-off weights from the airfields under study are then input into the *en-route* analysis, which will provide us with block times, block fuels and payloads, according to the conditions laid down. Aircraft performance analysis is not an exact science, so we should be prepared to investigate and evaluate special operating techniques that may colour the results.

Now we are armed with a set of data that will form the input for the economic analysis.

Chapter 6

Aircraft Economics

Why is Aircraft Economics so Important?

Unravelling aircraft economics has always tested the mettle of fleet planners. Its importance stems from the obvious fact that airline fleet decisions are being driven more and more by financiers rather than engineers. The huge investments made in aircraft and their support ensures that this is inevitable. Indeed, many airlines would put economics very high, if not at the top, of their list of key decision criteria.

There are many reasons why the study of economics is important, such as overall financial control of the business, benchmarking of a particular type of operation against others, and as an aid to pricing decisions. Our principal concern here is to use economic analysis to measure the effectiveness of an aircraft on either a route or network.

The reasons for the importance of aircraft economics are changing. The use of standard cost comparisons on single sectors has been relegated to a minor role in aircraft evaluations. In their place are more sophisticated modelling processes that take account of two key and more significant elements: revenues and the network effect.

The most appropriate way of addressing aircraft operating costs is to find ways not of minimising costs, but of managing them. There are often justifiable reasons for incurring higher costs if it can be shown that there is a more than compensating payback in terms of better reliability of operation or greater market opportunity. You reap what you sow.

In this Chapter we shall identify cost components, their measurement, and how they vary according to the type of operation. We will see the effect of aircraft type on their magnitude and examine the trends in cost evolution. Importantly, we will explore how to improve aircraft selection and assignment decisions by introducing revenues to the profit equation. This analysis will lead us into examining some useful modules to help understand aircraft positioning and profitability. Finally, we shall address one of the latest developments in economic thinking: dynamic capacity management.

Classification of Costs

A logical way to start an economic analysis is to design a structure upon which we can build up our knowledge of costs and revenues in such a way as to judge the impact of any change we may wish to make to the amount of capacity allocated. Any structure will, in all likelihood, be unique to one particular type of operation, in

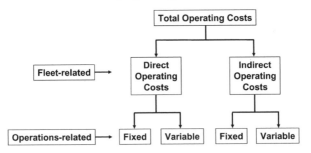

Figure 6.1   Classification of Operating Costs

one environment, and at one moment in time. Detailed comparisons between aircraft operating under different conditions are practically valueless. Published data can be useful to get a broad picture of where particular aircraft are positioned against each other, but there is no substitute for a customised analysis.

We must start by building our framework. The type of reporting seen in airline annual accounts is not appropriate for our needs because these data are designed to meet financial, rather than aircraft selection, criteria. Our main emphasis is on the operations side of the business. Non-operating costs can certainly be affected by the aircraft types in the fleet but we need to return to this complication when we build our investment appraisal later. Operating costs can be conveniently divided into two parts: Direct Operating Costs (DOC), being those costs which vary according to the type of aircraft used, and Indirect Operating Costs (IOC), which are those costs not affected by aircraft type.

Although this distinction sounds straightforward, there is continuing, and sometimes surprising, disagreement as to what constitutes a DOC or an IOC. One problem, as we shall see, is that certain costs can be allocated to a variety of functional areas, such as the aircraft itself, the network and the traffic.

A convenient way of visualising operating costs is seen in Figure 6.1, where we can sub-divide both the DOC and IOC portions into those parts which are fixed and those parts which are variable. A pure DOC/IOC division can be termed a fleet-related classification and the fixed/variable division is an operations-related classification.

Fleet-related classification   This is the simplest form of looking at operating costs. It is particularly useful for comparing the economics of aircraft on a given route without the added complications of network operation. This method is also used by manufacturers in aircraft design work and comparisons between competing types.

Despite, or perhaps owing to, its simplicity, a fleet-related classification can be a huge trap. There is no universally agreed definition as to what constitutes a DOC and any classification we use is arbitrary. Although there are a number of 'recommended' breakdowns, such as those issued by ICAO and the manufacturers, each airline should certainly build-up a customised DOC that is appropriate for its type of operation and network. We shall examine some points of disagreement as to what is, or is not, a DOC in a moment. In such a situation, we should be aware that economic calculations and even relative positioning may be driven by the make-up of the DOC, as well as the cost assumptions themselves.

Operations-related classification This is a far more useful way of looking at economics. By dividing costs into fixed and variable elements, we can measure the effectiveness of components of an operation. Once a value can be assigned to any adjustment of the operation we could use this information as an aid to pricing decisions.

As we work through the individual items in the DOC breakdown we shall see that aircraft utilisation plays a vital role in determining the fixed and variable division. Also, there may be significant differences between the degree of fixed and variable costs according to whether an airline is flying scheduled and charter services.

Cost Components

The following description of cost components covers the most usual elements of a DOC and IOC breakdown.

Fuel Costs (DOC)

The basic components of the fuel calculation comprise the price of fuel, the rate of consumption of the aircraft, the network operated, and the characteristics of the fuel itself.

The price of fuel   To a huge degree, airline management is hostage to movements in the overall price of oil, which are notoriously difficult to forecast. The dramatic rises in the price of crude oil in the 2000s was mirrored by rises in the price of aviation fuel. Indeed, the correlation between the two prices is almost 90%.

The effect of the fuel price increases has been to throw greater emphasis on the fuel efficiency of aircraft in fleet planning, as differences in fuel burn are obviously amplified. One way airlines can delay the effect of fuel price rises on their operating results is to hedge, or take an option of buying fuel on the futures market. Ironically, airlines that have judiciously protected themselves in this way have reaped the biggest benefits that they could ever have hoped to achieve through the acquisition of more fuel-efficient aircraft. One of the most difficult questions concerns whether fuel price is likely to remain at high levels for a considerable time. This may affect an airline's ability to raise capital to invest in fleet renewal at the most preferred time.

Apart from the underlying severe fluctuations in the price of crude oil, the 'price at the pump' is also subject to government taxation policies. Furthermore, individual airlines that are bulk users at an airport are able to negotiate discounts on the published prices, which again enhances their position against smaller competing carriers. Airlines that choose to serve airports where fuel prices are relatively higher are disadvantaged and may need to place more emphasis on the fuel efficiency of their fleet.

Another way in which airlines can exert some control over their fuel bill is to tanker fuel, as we saw in Chapter 5.

Aircraft fuel consumption   Clearly, this is primarily a function of the aircraft design. The percentage of the airline's costs associated with fuel will obviously diminish

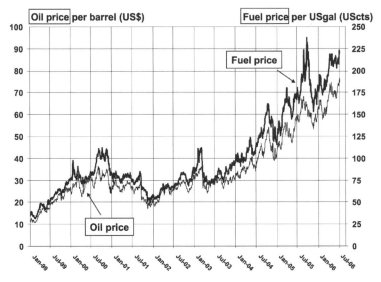

Figure 6.2   Evolution of Fuel and Oil Price
Source: US Dept of Energy

according to whether fuel-efficient aircraft are deployed. In an economic analysis it is important to decide whether fuel burns should reflect brand new aircraft or whether half-life conditions should be considered. In this case, some form of deterioration factor should be added to the initial calculation. The regulatory authorities sometimes impose a value, such as a 5% mark-up in the case of an ETOPS critical fuel calculation, unless a demonstrated value can be used.

Network design and operation   Selecting the most fuel-efficient aircraft is not necessarily the end of the story. Huge variations in fuel burn occur according to how the aircraft is used. As we have seen in Chapter 5, each aircraft design has an optimum payload and range characteristic. Operating in off-optimum conditions carries a penalty, even though network design may make this unavoidable. Aircraft are at their most efficient when cruising at altitude in steady-state conditions, but networks comprising short sectors inevitably mean that a greater proportion of an aircraft's life is spent in relatively inefficient phases of flight, such as take-off, climb, and low altitude manœuvres.

In dense airspace, a short-haul aircraft may spend a lot of time in holding patterns, or is perhaps only rarely able to attain optimum cruising altitudes. Another factor is the extent to which aircraft speed may need to be adjusted to make-up lost time or else ensure that connections are achieved at a hub. All of these factors will affect fuel burn, and therefore cost, and need to be considered for the economic calculation.

Apart from the flight profile itself, ambient conditions need to be fixed for the fuel calculation. We have already seen in our discussion of aircraft performance that the 'critical' case should be based upon a wind probability of 85%, measured on either a monthly, seasonal, or annual basis. The resultant fuel burn is thus only applicable to these slightly extreme cases. For the purposes of an economic calculation, however, it is important to adopt average values. So, the fuel burns must be re-run with 50% winds and an average, rather than maximum payload. Typical values for the economic fuel calculation would be:

Scheduled airline	75% passenger load factor
	50% cargo load factor
	50% wind probability for period under review
Charter airline	95% or 100% passenger load factor
	No cargo unless specified
	50% wind probability for period under review

Fuel characteristics   The measurement of fuel in an aircraft is based upon the volume of the tanks. It is therefore necessary to convert this value into a mass for the purposes of building up the weight of the aircraft. The most usual density used is 6.7 pounds, or 3.04kg, per US gallon.

Maintenance Costs (DOC)

Maintenance is an area which is probably the most difficult to tackle from the cost point of view, and the one which causes the most grey hairs on the heads of fleet planners. Maintenance costs are incurred during procedures to verify that airframe systems and the engines function correctly, to replace worn or defective components, and to manage the unscheduled failure of systems or components.

We can conveniently divide maintenance costs into Direct Maintenance Cost (DMC), covering labour and material costs associated with the maintenance of airframe components and engines, and Indirect Maintenance Cost (IMC), sometimes referred to as 'burden'. This covers overheads, administration, tooling, testing equipment and facilities, record keeping, supervision and quality control, and so on. Confusingly, these indirect costs are often merged with DMC as a part of the DOC breakdown, even though they are clearly not connected with aircraft type. Nevertheless, there may be *some* items that are aircraft-related, such as the maintenance of ground service equipment which may be specific to aircraft type. Also, aircraft painting may be considered as an IMC as it is image-driven, rather than technically driven. IMCs are very much a function of how a maintenance facility is organised and, as such, vary dramatically between airlines. For this reason, it is usual to account for IMCs as a percentage of the DMC. Values can range from 50% up to 200% of DMC.

DMC – line maintenance   These are the simplest tasks, such as checking tyre pressures and oil levels, which must be carried out either on a daily or weekly basis. As these tasks are calendar-related, the ratio of man-hours to flight hours is highly dependent upon the utilisation achieved. Also, some line maintenance tasks can be postponed until the aircraft returns to the main base, where more tools, spares and qualified manpower are available.

Airlines adopt very different approaches in clocking-on practice, which can radically affect line maintenance. In some airlines, mechanics clock-on when they

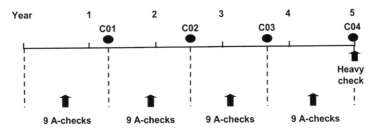

Figure 6.3   Block Maintenance

start to work on an aircraft, and clock-off when they complete their tasks. In other airlines, mechanics clock-on in the morning and clock-off in the evening allocating their whole working day to the aircraft, even though they may spend only an hour or so on the aircraft.

With line maintenance absorbing between 10% and 15% of DMC it is important to have a clear understanding of both maintenance and operational practice.

DMC – airframe maintenance   For each aircraft type the manufacturer issues a Maintenance Planning Document (MPD) which defines tasks and time intervals covering the airframe structure, systems and components. There will also be a component maintenance manual for items installed on the airframe by other equipment manufacturers, such as the auxiliary power unit and landing gear. In order to smooth the time intervals, the MPD groups together tasks into packages or blocks. The MPD reflects the manufacturer's recommendations for maintenance, which must then be built into an operator's customised maintenance programme, to be certified by the local airworthiness authority.

Airframe maintenance is composed of a number of checks, referred to as 'A', 'C' and heavy checks (referred to as the 'D' check for Boeing and McDonnell-Douglas aircraft). According to the type of aircraft 'A' checks can take place every 400 flight hours and 'C' checks every 15 months. With this rhythm it would be common for the eighth 'A' check of a cycle to coincide with a 'C' check. The heavy structural checks take place every five years, which may include painting, with additional structural checks every 10 years. Those tasks which are dimensioned by both flight hours and the calendar can be grouped into packages in order to avoid grounding the aircraft unnecessarily.

Figure 6.3 outlines a typical block maintenance timeline. We might find that the 'A' checks could be accomplished during night-stops, absorbing between 60 and 80 man-hours. The C01 and C03 checks might take 2.3 days each and the C02 check might take 3.3 days. The C04 check, combined with the heavy check, might take 21 days. In this example, therefore, the scheduled maintenance for a five-year cycle would require a total of 29 days.

However, it is possible to reduce the overall number of maintenance days over the five-year cycle by equalising the 'A' checks over the first three 'C' checks, as seen in Figure 6.4. The equalised 'A' checks are still performed during night stops, although the number of man-hours rises to between 65 and 90. The overall result of an equalised maintenance programme is that the total number of scheduled maintenance days over a five-year cycle falls from 29 to 21 days.

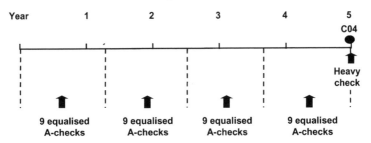

Figure 6.4   Equalised Maintenance

   Needless to say, if problems on the aircraft are found, such as corrosion or cracks, then additional work must be carried out according to the maintenance manuals. Not every task can fit into the regular system of 'A' and 'C' checks. For example, landing gear legs must be overhauled every eight years or 16,000 cycles, or landings, whichever comes first.

DMC – engine maintenance   This is a function of both operating hours as well as cycles, or take-offs. Every time an aircraft takes off the engines are running at nearly their maximum power output, causing wear and thermal stress. Furthermore, operating conditions such as hot and high-altitude airfields and short runways, also contribute to wear.

   The days of complete engine overhauls have been largely superseded by modular overhauls. So the first engine overhaul is not likely to be expensive, as not all modules would need replacing. However, engines do comprise a number of expensive Life Limited Parts (LLPs) which must be scrapped after their specified life limits have been reached.

   Figure 6.5 shows an approximate breakdown of maintenance costs for an A320.

Factors Influencing Maintenance Costs

These fall into four principle categories: airline, aircraft, geography and route network.

Airline influence on maintenance costs   The degree to which an airline sub-contracts or undertakes its own maintenance determines the degree of investment made in

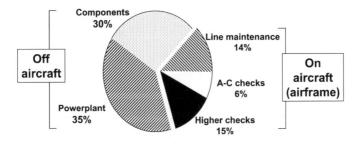

Figure 6.5   Maintenance Cost Breakdown – Example

facilities and the labour costs incurred. A maintenance policy should embrace all aspects of the organisation of the activity, from training, the extent to which tasks can be equalised, up to the investment in tooling and spare parts. It is a huge subject.

One area that needs attention is the extent to which Service Bulletins (SBs) are incorporated on the aircraft. Not all SBs are mandatory and investment can be saved by only implementing those that are essential to the operation. Airlines committed to ETOPS find that some SBs are mandatory, for example. In any event, the future value of the aircraft could be damaged if SBs are not implemented.

An aircraft selection process could be easily and radically affected according to the style of maintenance activity of the airline.

Aircraft influence on maintenance costs   It is relatively easy to pinpoint design and operational advantages of specific aircraft types. Aircraft coming off the production line today boast various forms of on-board maintenance surveillance, which progressively reduce the need to remove suspect components for test, only to find that they are not defective after all. Airlines building up a family of aircraft of the same technical standard undoubtedly benefit from synergies that come from technical knowledge and spares commonality. Care must be taken, however, in making sure that large discrepancies in the rank of the aircraft in the production line of the same aircraft type do not mean that production standards have evolved to such an extent as to erode commonality benefits.

When comparing aircraft that have similar take-off weights the amount of thrust available is also a consideration. Higher thrust engines carry weight and drag penalties due to their size, and twin-engine aircraft designed for extremely long missions are generally penalised due to the need to ensure that sufficient thrust remains for the engine-out case on take-off. On the other hand, a surplus of thrust also means that more opportunities exist for de-rating the engine for specific take-offs, which translates into less wear on the engine and, ultimately, lower engine maintenance costs.

The ageing of an aircraft will affect baseline maintenance calculations in two areas. Firstly, new aircraft enjoy a honeymoon period when actual DMCs are below predicted mature levels. This is due to a combination of the newness effect, particularly of engines, and secondly the effect of warranties on components. For a new type in the fleet these advantages may need to be balanced by additional costs as mechanics proceed down the learning curve. However, in general, aircraft maintenance will be lower than the mature levels for up to four or five years after entry into service.

The second ageing effect is, unsurprisingly, where more elderly aircraft begin to show signs of incurring higher than mature levels. In general, maintenance costs start to rise from around 12 to 14 years beyond entry into service.

During the course of the life of the aircraft, as we have seen, there are periodical heavy checks which take the aircraft out of service for extended periods and which can cost several million dollars. The amount of work required is largely contingent upon the specific condition of the aircraft, so it is difficult to budget precisely. One common approach is to make a budgetary allowance on an annual basis to avoid heavy spikes of expenditure. The manufacturer is thus able to provide maintenance cost estimates either with or without a provision for heavy checks. Figure 6.6 illustrates the principle.

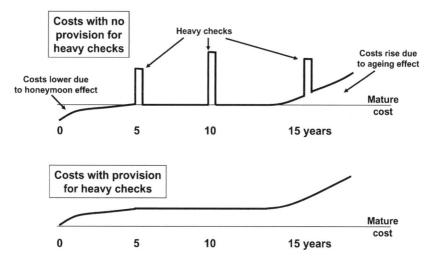

Figure 6.6   Maintenance Costs – Ageing Effect

Geographical influence on maintenance costs   Operators based in remote locations must carry the additional cost of being far from a supply of spares. It is not just a question of money, but also of time. Geography may dictate the degree to which maintenance can be performed locally or back at home base. Remoteness and skilled and experienced maintenance personnel do not necessarily go hand-in-hand.

Challenging climates also take their toll on the maintenance of an aircraft. Desert operations result in increased damage from sand ingestion, and coastal operations are plagued with corrosion issues.

In some environments, such as India and China for example, taxiways and runways are prone to be contaminated with debris with the obvious risk of engine ingestion.

Route network and operations   This is by far the biggest influencing factor among the four categories. The structure of the network will to a large degree determine the utilisation that can be achieved from the fleet. Clearly, short-haul scheduled operators have a somewhat greater challenge in maximising their utilisation than long-haul operators. The effect on maintenance cost can be dramatic. It is not unusual for the same aircraft type to incur double the DMC on a short-haul compared to a long-haul network. Indeed, short-haul operators might find that maintenance cost overtakes fuel as being the prime cost-driver, whereas a long-haul airline could find the reverse.

Any calendar-related checks, such as daily and weekly checks, consume virtually the same number of man-hours irrespective of utilisation. The effect is that the man-hours per flight hour would increase threefold if daily utilisation reduced from eight to two hours/day.

Linked with utilisation is the average sector length. Every take-off and landing, or cycle, involves the acceleration and deceleration of engines, and use of the landing gear and lift control devices such as slats and flaps. Also, the fuselage is

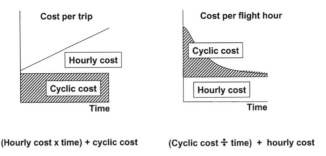

(Hourly cost x time) + cyclic cost   **(Cyclic cost ÷ time) + hourly cost**

Figure 6.7   Maintenance Costs – Influence of Sector Length

pressurised and depressurised, and the doors closed and opened. Many cyclic-related costs involve expensive items, such as wheels, tyres and brakes. Higher average sector lengths tend to mean that the aircraft performs proportionately fewer cycles in relation to the hours flown. Figure 6.7 shows this relationship.

The structure of the network may also dimension the number of occasions each aircraft can return to main base for maintenance attention. So, long-haul linear-style operations may require more maintenance being undertaken by third parties at outstations, for example.

Maintenance Cost Modelling Parameters

Owing to its complexity, the modelling of maintenance cost is undertaken by specialists. There is not a universal method of determining the costs as so much depends upon individual cases. The most usual analytical method is based upon the ATA chapter system, from which hourly and cyclic elements can be derived for both labour and material costs.

Data sources to enable the modelling of maintenance costs are many and varied. Apart from manufacturers' own data, IATA produce a standard set of data for comparison based upon airlines contributing to the Production Performance Measurement working group. These data are adjusted for common assumptions to enable meaningful comparisons to be made for similar aircraft types. Other data sources include audits undertaken with individual operators.

Customisation of DMCs is naturally dependent upon accurate input data, such as labour costs for both in-house and sub-contracted work.

An important set of parameters concerns labour efficiency, plus material handling and usage factors. A 100% labour efficiency factor would be where all tools are readily available to perform a task and where no time would be wasted by the mechanics. In practice a 75% labour efficiency is presumed for an efficient airline. Material handling comprises an additional charge to take account of the packaging and shipping of materials. For in-house work this would be zero, rising to up to 25% for sub-contract work. Lastly, the material usage factor takes account of those parts that are scrapped due to shelf-life expiry or else accidental damage. A factor of 5% would be typical. Figure 6.8 illustrates a typical set of assumptions required for a customised DMC study.

Network assumptions:

Average sector (hours):	4	
Annual utilisation (block hours):	3 000	
Taxi-time (minutes):	15	

> **Typical assumptions needed for customised study for a long-haul operator sub-contracting engine maintenance and higher checks**

Economic assumptions:

Base year for study:	2007	
Currency and exchange rate:	US$; 1	

	In-house	Sub-contracted
Labour rate	$70	$90
Labour efficiency	65%	75%
Handling	0%	15%
Material usage efficiency	95%	95%

Operator activity:

Line maintenance	100%	Component overhaul	30%
A-C checks	100%	Powerplant systems	0%
Higher checks	0%	Engines	0%

(Burden and Life-Limited Parts excluded)

Figure 6.8   Maintenance Cost Assumptions

Maintenance Reserves

Our discussion so far has concentrated on modelling of maintenance costs for owned aircraft. In the case of a leased aircraft it is the lessor, as the owner, who bears the brunt of maintenance costs, especially the heavy checks. The lessor therefore requires a cash reserve to be built up to ensure that enough money is 'in the pot' when maintenance tasks are due. This system ensures that the lessee pays his share of the heavy maintenance tasks, even if the lease terminates before the work falls due. When a second lease becomes effective, the 'opening balance' will reflect the amount of reserves built up under the first lease. There is always the risk that maintenance reserves turn out to be higher than the actual cost of performing a heavy check. Should the reverse be true, then the lessee would be expected to contribute the shortfall.

Maintenance reserves usually cover four main areas: airframe, landing gear, auxiliary power unit (APU), and engines. Reserves do not include the costs of line checks, 'A' or small 'C' checks. However, the more costly five and 10-year checks must be accounted for.

Reserves are usually paid monthly, the calculation being based upon an agreement concerning the number of hours to be flown. The actual payments are made according to the hours performed for the previous month.

Maintenance Cost Third-Party Agreements

Even airlines that own their aircraft sometimes prefer to enter into an agreement with an independent maintenance organisation whereby, in exchange for a monthly fee, a total maintenance package for the engine is provided. There is certainly a trend in this direction, brought on by the tendency for airlines to focus more on their core business of transporting passengers, rather than become involved in what they see as peripheral support activities. Some large airlines, like Lufthansa, have spun-off their maintenance, repair and overhaul organisation and created a separate business unit. Such organisations have the ability to offer packages that encompass not only the engine but also the airframe.

A huge advantage in contracting out maintenance in this way is that maintenance costs are predictable. Also, the investment normally required by the operator to support the maintenance activity is avoided so that capital can be deployed elsewhere. Another advantage is that the maintenance program can be customised for the aircraft as a function of the operation. Thus, low-cost carriers need their maintenance performed at night, but charter operators do not have the luxury of downtime at night.

There are some potential disadvantages in contracting out, too. There is always the risk that a large maintenance provider will assign lower priorities to small clients, who may need to accept compromises in the timing of work and even where the work is performed. This suggests that the operator may suffer a loss of control and flexibility. Also, the high reliability of today's engines means that heavy expenditure on a new aircraft is unlikely to be incurred in the early years of an aircraft life. However, regular monthly payments will still need to be made, only to accrue in the accounts of the maintenance provider.

It is important that a contract is negotiated to protect all the parties involved. One of the largest contracts for a single aircraft type took over one year to negotiate. This was a 10-year $1 billion contract between easyJet and Zürich-based SR Technics to maintain their entire fleet of A319s.

Crew Costs (DOC, and Sometimes IOC)

Flight crew costs are clearly a DOC but cabin crew costs are considered by some, such as ICAO, to be indirect costs. The rationale for this is that the cabin crews are involved with the passengers and therefore are considered as traffic, rather than aircraft, dependent. Most fleet planners are happy to include cabin crew costs as direct costs, however, as the number of cabin crew varies according to the size of the aircraft and it is not logical to divide up the personnel on the aircraft into different categories of operating cost. All personnel on an aircraft are involved in its operation.

Flight crew (DOC)   There are three components: salaries, allowances and training. For the purposes of fleet planning we need an annual salary value, or else hourly cost, including social charges. There are huge variations between airlines of pay scales and seniority conditions, of course. Allowances are those expenses incurred during time spent away from main base. Training costs can be usefully split into initial training and recurrent training, with large savings possible where common flight decks are present in differently sized aircraft in the fleet. Recurrent training can either be incorporated into the annual salary value, or else accounted separately as an hourly value.

Flight crew productivity   Modern large aircraft require two flight crew members. However, additional flight crew is needed for very long flights of around 12 hours or more. This threshold is determined by regulators and may also be driven by airline policy and union issues. Where additional flight crew members are to be considered, it should be remembered to include their weight, plus their baggage, in the operator's items in the weight build-up.

The amount of flying hours performed is often measured on a monthly basis, with values such as 70 block hours per month being typical. This means that several crews would be needed per aircraft in the fleet. If we know that the utilisation of the aircraft is 4,500 hours, we can deduce the number of crews from the monthly number of flying hours per crew member. In this case it would be 5.35 crew members.

Network shape can be a huge determinant of flight crew costs. Airlines that operate a short-haul service, with all aircraft returning to the main base at night, incur relatively lower costs than long-haul operators flying around the globe. It has been known for 10-day layovers to occur where loose scheduling precludes the recovery of a crew by another aircraft.

Cabin crew (DOC or IOC)   The cost components are similar to those of flight crew, except that recurrent training costs are not significant. Again, the costs may be provided in either an annual salary form, or else as a value per block hour performed.

We have to place particular attention to the number of cabin crew assigned to a particular aircraft type. Regulations dictate the minimum number of cabin crew, which is one crew member for 50 passenger seats. However, airline service standards and sector length dictate actual practice, with ratios rising as high as one crew member for four passengers in First-class. We completed this assessment when we built up the interior configuration of the aircraft and fixed the number of attendant seats in the layout. It is important to crosscheck that the economic calculations mirror the actual layout in the evaluation.

Typical values are as follows:

One cabin crew member per: 4–15   First-class seats
                               10–20   Business-class seats
                               20–50   Economy-class seats

Regulations aside, there is a dangerous trap that a fleet planner can fall into when comparing the economics of aircraft with slightly different seating configurations. If we are comparing two single-class aircraft with a seat row's difference between them, with one layout having 118 seats and the other 124 seats, and the cabin crew ratio is one attendant for every 20 seats, then the evaluation rules would suggest that the smaller aircraft requires six crews and the larger aircraft requires seven crews. Obviously, there has to be some point at which an increase in seats triggers another crew member, but one might wish to presume that the same number of cabin crew would operate both layouts.

Low-cost carriers consider this to be a critical issue and will dimension the configuration of their aircraft to maximise the productivity of cabin crew.

Landing Fees (DOC)

This DOC is usually divided into two components: those fees which are aircraft dependent fees; and those related to the traffic.

Aircraft dependent fees   Mostly, charges are levied according to the registered take-off weight of the aircraft. One exception to this is the US, where the Maximum Landing Weight can be found as a charging parameter. Operators are not completely hostage to airport landing charges because it is possible to certificate an aircraft at design weights lower than the MDTOW in order to reduce the impact of landing fees. However, this policy can only be pursued so long as the necessary payloads can be carried on the network. If there is a significant excess of design weight over that required for the operation, perhaps there is a deeper problem in that the wrong aircraft configuration is being used.

Apart from a basic landing fee, there is often a multitude of supplementary charges to cover the use of the ramp, parking areas, security, local air traffic control fees and even airport lighting.

Airport authorities can impose peak charging policies in an effort to encourage users to switch their landing times to off-peak periods. When this happens airlines rarely react by changing their scheduling, suggesting that the costs of uprooting the operation would far outweigh the extra landing costs they incurred.

Traffic dependent costs   The second element of landing fees is that related to the number of passengers carried. These charges take into account the use of the terminal and security facilities. In fleet planning, these charges are not usually considered because they are presumed to be levied as part of the ticket price.

The most critical area for fleet planning is to identify where expensive airports are likely to have a greater impact on those aircraft that have relatively higher take-off weights.

Navigation Fees (DOC)

These fees, also a DOC, are over-flight fees paid to each state over-flown on a mission. They are based upon the Flight Information Regions (FIRs) and the flight plans filed, which may not actually represent the actual routeings flown. It may sometimes be the case that the most optimum routeing in terms of fuel burn might not be the most optimum for navigation charges if that routeing happens to 'clip' the corner of a particularly expensive FIR. In such a case, a small routeing deviation might contribute an overall cost saving.

There are, as with most cost items, big differences by region. In the USA there are currently no navigation fees levied. Some jurisdictions charge a flat fee for over-flight but the majority add a charge associated with distance flown and take-off weight. The most common formula is that of Eurocontrol, widely applied in Europe and elsewhere:

$$\text{Unit rate of country} \times \text{distance} \times \sqrt{(\text{MDTOW(tonnes)}/50)}$$

Insurance Costs (DOC)

This DOC is usually divided into several components: hull insurance, war and political risk, deductible insurance (similar to a franchise), and third party liability insurance, which is mandatory. Factors influencing the hull insurance include the airline safety record, aircraft price, fleet size and age. For fleet planning purposes the premium is linked to the price of the aircraft and remains constant, being prorated according to block time. For a second-hand aircraft the premium is linked to the fair market value. The amount paid annually can vary between 0.5% up to 5% of aircraft price.

Ownership Costs (DOC)

Here we have an area of some contention, as there is opportunity to alter the shape of the DOC breakdown without actually changing the true economics of the aircraft under review.

Depreciation   The purpose of depreciation is to allocate the initial cost of the investment over a period of time. Unlike every other item in the DOC breakdown, depreciation is not a cash cost and no money leaves the business. Its purpose is simply to reflect the value of the aircraft in the books at any moment in time.

The most usual method of depreciating aircraft is the simple straight-line method. There are two key decisions that must be made. Both the depreciation period and residual value of the aircraft must be fixed. These two elements allow us to easily calculate an hourly depreciation charge as follows:

$100m investment depreciated over 15 years to 10% residual value
and 3,500 hours utilisation *per annum*
($90m/15)/3 500 = $1,714 per hour

It would be an easy matter to reduce the hourly depreciation charge, either by elongating the depreciation period, or else raising the residual value percentage. An airline might wish to do this in order to reduce the operating cost, thereby improving profits declared. However, accounting rules preclude too much tampering with these parameters. It can, and has, been done, but generally such adjustments can only be made once. On the one hand it might seem logical to extend the rather traditional 15-year period of depreciation in order to reflect the growing longevity of today's aircraft. Likewise, why not push up the residual value, too? There are risks associated with this practice. Firstly, maintaining the aircraft on the accounts for extended periods simply burdens the business with a liability that may be increasingly less viable toward the end of its life. Secondly, if the forecast residual value cannot be realised then the airline might have to incur a book loss on the aircraft at the end of its life.

Under some tax jurisdictions it might even be more appropriate to reduce the depreciation period. Such a policy of accelerated depreciation has been successfully employed by Singapore Airlines, who have been able to reduce tax on their profits. Coupled with this advantage, Singapore Airlines built up a fleet of quite new aircraft that had been fully depreciated and no longer on their books.

So, in calculating depreciation charges in the DOC, we should heed the dangers of manipulating comparisons by artificial means.

It is not unusual for an airline to depreciate different aircraft types over different periods, according to the useful life of the asset. For example, Austrian Airlines depreciates its regional jets over as little as 12 years, single-aisle main line aircraft over 18 years, and its long-haul fleet over 20 years. Most airlines fix their depreciation period for aircraft between 15 and 20 years.

Loan payments and other acquisition costs   Another area of contention concerning the assessment of ownership in the DOC is whether or not to include interest charges.

There is a school of thought that says that any loans taken out to acquire the aircraft are a reflection of the financing needs of the airline and not the efficiency of the aircraft, so loan repayments should not be included in the operating costs as they serve to dilute the true assessment of the aircraft economics. Indeed, the aircraft manufacturers should not compete with each other on financing terms. Also, interest payments are strictly a non-operating expense and do not appear on the profit and loss statement.

The opposing view is that interest payments are a fact of life and eliminating them from the DOC means that not all costs associated with the acquisition of the aircraft are being considered.

Where things get tricky is when we might wish to compare a purchased aircraft with one on an operating lease. In this case, the operating lease payment would certainly reflect the need for the lessor to pay interest for *his* purchase of the aircraft. This being the case, then the directly purchased aircraft should also be burdened with interest payments.

There is no right or wrong here. However, we do need to be aware that interest payments do dilute the effect of other costs that may compromise any comparisons. Also, there is no relationship between the true economic capability of an aircraft and the loan conditions.

Airlines may fix interest rates up to three years before delivery in order to take advantage of constantly evolving market rates. In addition, pre-delivery payments can also be financed, so interest payments can start before the delivery of the aircraft.

In export credit financing there are two basic types of financing: a loan structure and the more commonly applied lease structure. In the case of a loan, the export credit insurers give the loan directly to the airline together with the title of the aircraft. In a lease structure the export credit agencies give the money to a Special Purpose Company (SPC), which holds title to the aircraft and pays the manufacturer. The airline pays rentals to the SPC and gains title when the entire loan has been paid.

Under a loan structure the repayments are made with constant principal and declining interest, whereas under a lease structure the repayments are of a mortgage style, with constant payments. Figure 6.9 shows the two types of payment in diagrammatic form.

The disadvantage of the loan structure is that it places a burden on the airline cash flow in the early years. The constant payment stream of the lease structure is often the preferred option.

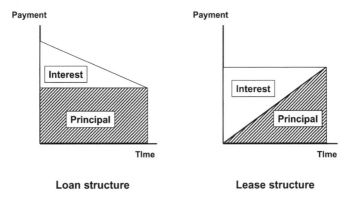

Loan structure   **Lease structure**

Figure 6.9   Interest Payment Options

The oddity about the treatment of ownership costs in the DOC is that depreciation is a non-cash operating cost that is included in the breakdown, whereas interest charges are a cash non-operating cost and often excluded. However, if interest charges are excluded from the DOC, this is merely to say that they are accounted for as an entity separate from the DOC itself.

Leasing charges to a lessor are most certainly included as a DOC. It is useful to separate the rental charge from maintenance reserves.

Handling Costs (DOC and IOC)

Now we begin to enter the realm of Indirect Operating Costs, which are those costs which are not affected by the type of aircraft operated. Confusingly, handling costs are partly direct and partly indirect.

Aircraft handling involves ground manoeuvres at the airport, aircraft servicing, flight administration and the amortisation of aircraft-specific handling equipment, such as airstairs and cargo loading equipment. Traffic handling involves passenger check-in facilities, boarding and de-planing, baggage and cargo handling. The level of costs is a function of the equipment present at a particular airport, labour rates and efficiency, the competitive situation and the operating patterns. For example, it becomes more expensive to support an aircraft type that serves a destination on an infrequent basis, especially if there are no other operators of the type in the vicinity. A cost comparison between two competing aircraft types would need to reflect the overall handling environment.

One particularly important area for study is that of baggage and cargo handling with containers. Operators of single-aisle aircraft who adopt a form of containerisation find numerous advantages in terms of speed, less damage and loss, interline capability, and less industrial injury claims due to the reduced need for handlers to clamber in and out of confined spaces in the underfloor. These advantages need to be weighed against the cost of providing and maintaining the containers that also affect payload owing to their additional weight in the aircraft. However, the benefits are very apparent for short-haul airlines needing to achieve rapid turn-rounds and maintain a good market image.

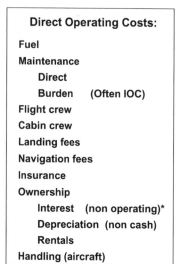
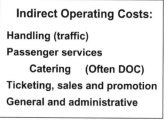
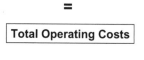

Direct Operating Costs:

Fuel
Maintenance
   Direct
   Burden   (Often IOC)
Flight crew
Cabin crew
Landing fees
Navigation fees
Insurance
Ownership
   Interest   (non operating)*
   Depreciation (non cash)
   Rentals
Handling (aircraft)

Indirect Operating Costs:

Handling (traffic)
Passenger services
   Catering   (Often DOC)
Ticketing, sales and promotion
General and administrative

=

| Total Operating Costs |

*Not considered as a DOC in IATA,
ICAO, AEA, AAPA breakdowns

Figure 6.10   DOC and IOC – A Fleet Breakdown

Handling cost modelling would normally be related to the payload and size of aircraft.

Passenger Service Costs (Sometimes DOC, and IOC)

Yet again we find a cost which can be considered to be partly direct and partly indirect. The DOC element is usually the catering costs, including all food, beverages, amenities and give-aways. One argument for including catering costs as an IOC is that the number of meals prepared is a function of the total amount of passengers transported, and is not contingent upon the composition of the fleet. On the other hand, to support the idea that catering could be a DOC, larger aircraft do bear a higher allocated catering charge because they carry more passengers. On-board sales of duty-free goods and other products generate revenues that can also be incorporated into the DOC breakdown. Charter operators and low-cost carriers find that significant revenues can be generated in this way.

Passenger service costs also involve ground service, such as the provision of lounges, and are truly indirect with no impact at all on the type or size of aircraft operated. Modelling can be a fixed rate per RPK or RTK.

Ticketing, Sales and Promotion Costs (IOC)

Once more, these costs are totally indirect, comprising sales costs, commissions to agents, computer reservation system fees, and all advertising and promotion costs. Commercial cost modelling is usually related to the revenue generated. This is one cost area that has come under attack. The growth of Internet usage has resulted in a high level of pricing transparency for airline customers and consequently the power of the travel agent has diminished. For airlines with a relatively simple pricing structure,

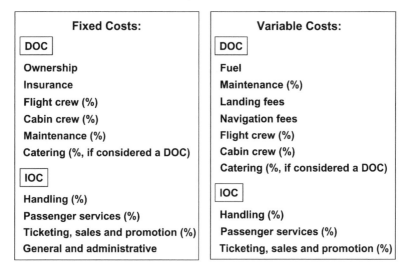

Figure 6.11 DOC and IOC – An Operations Breakdown

and this includes most low-cost carriers, it has been possible to significantly reduce this particular cost category by migrating distribution onto the Internet.

General and Administration Costs (IOC)

Our last IOC item involves all overhead costs such as management and corporate expenditure, public relations costs, and buildings and equipment that are not aircraft-related. To assess these costs one has to decide the degree to which they are actually production-related, in which case they can be calculated according to ASKs or ATKs generated.

Grouping the Cost Items

Having listed all of the items that generally make up the operating cost we can now summarise them, in Figures 6.10 and 6.11 from the perspective of a fleet-related breakdown.

A variation on the fleet-related breakdown would be to completely exclude ownership costs in order to focus more clearly on the economic performance of the aircraft. Such a cost division is termed *Cash* Operating Costs, or COC.

When it comes to defining an operations-related breakdown, however, we will often find that there are difficulties in deciding the degree to which some costs are either fixed, or variable. Crew costs are easily dealt with as salaries are fixed and allowances are variable. Also, maintenance is relatively easy to split. However, there are always problems of allocation when it comes to handling and commercial costs.

If the allocation problems can be resolved then this operating-related view of DOCs is extremely useful in determining the true value of actually operating, rather than just possessing the fleet. Factors that drive the division include load factor, utilisation, sector length and network shape.

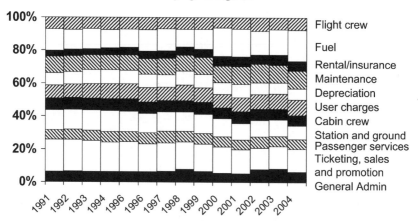

Figure 6.12   Evolution of Operating Costs
Source: IATA Economic Results and Prospects

Evolution of Operating Costs

In the beginning of the 1980s the largest single element of the total operating cost breakdown was fuel, accounting for 25% of costs. Part of the reason was the relatively higher price of fuel due to political effects and another part was that airline fleets were still populated with aircraft types which were considerably less efficient than those on offer today. As airlines undertook huge re-equipment programmes with more fuel-efficient aircraft, the cost breakdown evolved so that fuel represented only around 11% of costs, with ticketing, sales and promotion costs becoming the biggest cost. That situation has now reversed, due to the significant fuel price increases of the 2000s. However, it is too easy to generalise and forget that airlines incur significantly different cost structures according to the type of operation they pursue, their geographical location and their route structure. In fleet planning, we must pay heed to the factors which drive costs and judge the extent to which the aircraft can play a role in their management.

Figure 6.12 shows the evolution of total operating costs in recent years. Note that IATA use slight variations in terminology to that used above. For example, 'station and ground' equates to handling charges. We can detect several trends in this generalised view. Firstly, general and administration costs are gradually rising. Secondly, the industry does seem to have a better grip on those very significant ticketing, sales and promotion costs. Another trend is for rentals to assume a large percentage of the total as more airlines turn to leasing as an acquisition alternative. The largest items in the DOC side of the equation are maintenance and user charges. The former is to a large degree under the control of airline management, whereas airlines are rather more hostage to the payment of landing and navigation fees, according to their choice of network.

The overall trends in economic behaviour are worth watching because they give clues as to how the big-ticket items are developing on a global basis. These trends can be consulted in the annual 'World Air Transport Statistics' of IATA or else the ICAO financial reports. In addition, IATA publishes 'Economic Results and Prospects' annually,

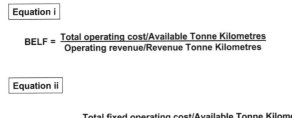

Figure 6.13 Break-Even Load Factor

which tracks movements in operating costs. However, not all data includes all airlines. For example, major low-cost carriers are not IATA members, so their dramatically lower ticketing, sales and promotions costs do not appear in the IATA data.

Although a global vision of airline costs is worthwhile, we must relate costs to a specific operation in order to be able to make a correct judgement concerning the impact of the aircraft.

Methods of Cost and Revenue Measurement

Break-Even Load Factor

The most fundamental measure of operating performance is the break-even load factor, or 'BELF'. This is the percentage of capacity filled so that the total costs incurred exactly equal the revenue generated. The BELF can be assessed at company level, where fully allocated costs are set against all operating revenues, or else at an individual sector level, where perhaps only a proportion of costs are considered. Where the BELF has been considered only on the basis of DOCs, then much lower values will clearly result.

There are two ways of calculating the BELF. A simplified approach is to take all costs together (Figure 6.13 Equation i), whereas a more appropriate method is to relate fixed costs to production and variable costs to the revenue (Equation ii).

We should not read too much into individual sectors that do not reach their break-even point. Such routes may contribute important connecting traffic to other parts of the system, which might be poorer if loss-makers were eliminated.

Operating Leverage

Presuming that we can accurately determine our fixed and variable costs then it should be straightforward to assess the overall operating leverage based upon the total amount of activity performed. As total activity expands there will be occasions when fixed costs will suddenly rise as specific investments are made. Apart from major fleet acquisitions, these may include the costs of opening new stations, investing in a new simulator, additional spare engines, or else expanding the size of administrative accommodation. With the possible exception of route expansion, such one-off investments will not necessarily interfere with a progressive increase

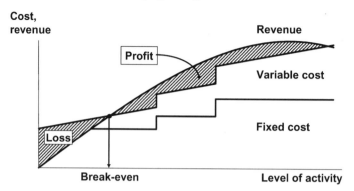

Figure 6.14 Operating Leverage

in revenue. Figure 6.14 shows that there may be levels of activity where profitability can be optimised. If the operation is really overstretched, with perhaps too much capacity having been introduced, then it is even possible for the revenue curve to begin to invert, where markets become so competitive and price-elastic that discounting cannot sustain revenue increases.

Cost Escapability

A popular method of measuring costs is to consider them in terms of escapability. Knowledge of the fixed and variable proportions of costs is essential. This type of assessment is simply an identification of which costs are completely avoided if part of the operation is dispensed with.

Different levels of escapability are appropriate according to the depth of analysis. For example, a short-term planning approach would be to identify costs that could be avoided if a single flight did not operate. Fuel, some maintenance costs, landing and navigation fees, catering, handling and any crew allowances could be avoided. Revenues set against these costs could be seen as a first line of contribution to the system. We should always remember that cancelling a service to save costs is all very fine, but revenues are being lost which may stem from connecting traffic in other parts of the network.

The medium-term planning approach, at the level of the schedule, would examine route-related costs, the number of aircraft in the fleet and crew salaries, for example. All variable costs and some fixed costs, like those associated with maintenance work, could be included at this contribution level.

Finally, for long-term investment purposes, all costs could be included, such as aircraft ownership, spares, simulators and all overheads.

The main virtue of looking at cost escapability is for budgeting and accounting purposes. Certainly, it is useful to identify whether individual services or aircraft are making a contribution at various levels to the business. However, there are risks in fully allocating costs to a schedule. If, for example, half the routes in the network generate a margin of 10% and the other half generate an equivalent loss, then the entire system is breaking even. However, eliminating the loss-making routes would simply burden the profitable part of the network with all of the overheads, risking the whole system.

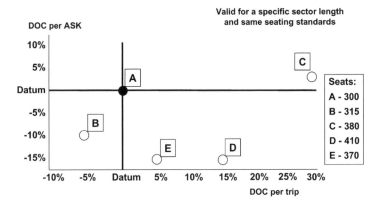

Figure 6.15   The Fan Diagram

Fan Diagrams

A common method of displaying costs is by the 'fan' diagram, such as in Figure 6.15, which focuses on unit cost.

At the end of the 1960s Boeing used the concept of unit cost in order to demonstrate the economics of the soon-to-be-introduced 747. This aircraft represented a huge leap in size compared to the aircraft it would replace, such as the DC-8 and 707. Dividing the operating cost by the number of seats on the aircraft was a logical proof of efficiency. The unit cost concept has been with us ever since, often expanded to embrace the underfloor capacity as well. Multiplying the unit cost by distance flown does give us a useful parameter for cost measurement. Cost per Available Seat-Kilometre (ASK) and Available Tonne-Kilometre (ATK) accentuate the economies of scale of large aircraft quite successfully.

The fan diagram enables us to easily position aircraft economically. From Figure 6.15, if we take the 300-seater Aircraft A as our datum, competing products might radiate in any of the four quadrants according to whether they are better or worse in either trip cost or unit cost. The unitary measure in the example is seats, but the total airframe capacity measured in tonnes might also be applied.

Aircraft B is better than Aircraft A in both unit and trip cost, even though it is slightly larger. Aircraft C, on the other hand, carries around a 30% trip cost penalty owing to its much larger size, but is only slightly more expensive in unit cost. Aircraft D and E are both 15% less expensive than the datum, the difference being partly explained by the different sizes of these two products. Although Aircraft C and D are fairly close in size, they are very far apart in economics. With no other information to hand, Aircraft B might appear to be the most attractive.

Problems of fan diagrams   However, we are missing quite vital information in three key areas: revenue, technology level and the network effect.

Fan diagrams based upon unit and trip costs say nothing about one-half of the economic equation: revenue. As aircraft size changes, so do their revenue-earning abilities. The fan diagram can tell us nothing about whether the market addressed by our datum aircraft could alternatively be served by Aircraft C or D. Also, a 380-seater market would not be served in the same way as the 300-seater market. For a start, frequencies offered would need to be different, which might itself impose a different spread of yields between the competing types. Differently sized aircraft not only earn different amounts of revenue, but also carry different degrees of risk as well. Much care is needed in interpreting cost data without considering a corresponding revenue assessment.

The second problem of the fan diagram is that, rather obviously, it can tell us nothing about the standard of technology in different aircraft. Perhaps we would expect Aircraft D to have poor economic performance compared with Aircraft C if there is a big difference in technology. However, if both aircraft were of a similar generation of technology then our perspective on the comparison should change. If Aircraft C is of an earlier technology standard than our datum aircraft, then the economic positioning is a powerful argument to reduce aircraft size, especially if a frequency advantage could be exploited. Although there is undoubtedly a sizeable market for very large aircraft there are still opportunities to play the frequency game as markets continue to liberalise.

Thirdly, fan diagrams tend to be based on calculations for a single sector length. Each aircraft design has its own optimum range, based upon a certain payload. Selecting a generic sector for comparison will inevitably mean that one aircraft of a pair is likely to be closer to its true optimum. Although switching the sector length for the fan diagram comparison is not likely to switch the ranking of aircraft, at least we should be aware of the effect.

The issue of sector length in this type of comparison is especially significant because aircraft may be compared with quite different range capabilities. This would suggest that an airline would perhaps deploy aircraft differently, too. Different deployment means that utilisation would differ and that, in turn, completely upsets any meaningful comparison because fixed costs would be distributed differently.

Despite the above problems fan diagrams are valuable for positioning of aircraft, although we cannot base decisions upon them. As usual, we need to be very aware of the assumptions that have been applied in order to properly interpret the results.

Fan diagram modelling and the schedule   Whenever we construct a cost comparison, it is very likely that a schedule will already exist, based around an existing network. This necessarily means that the economic assessment of the aircraft is being dimensioned around the scheduling environment, however perfect or imperfect that may be.

A feature of aircraft economics is that as stage length increases, unit costs improve. This is simply because fixed costs become more widely spread and aircraft are more efficient when cruising than when manoeuvring at low speed and low altitude.

Imagine a simple route network upon which the differently sized Aircraft A and C are being compared, as shown in Figure 6.16. Owing to its large size the 380-seater might only be able to operate from XXX to YYY four times a week, with an onward 'tag-end' to ZZZ. However, the smaller 200-seater would enable both YYY and ZZZ to be served directly on four occasions a week, plus two 'tagged' services. The smaller aircraft has enabled a 50% increase in frequencies, and more seats are

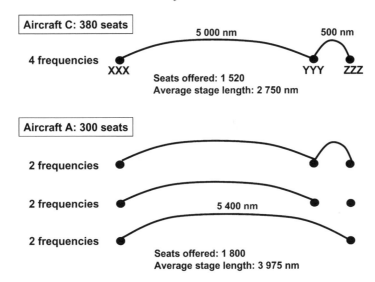

Figure 6.16   Multi-Leg Cost Modelling

available overall to take advantage of market stimulation due to this effect. However, the average sector length of this mini-network has increased from 2,750nm with Aircraft C to 3,975nm with the Aircraft A operation.

So, the increase in average stage length has fundamentally shifted the fan diagram positioning of the two aircraft. If the effect of the higher stage length were to diminish unit cost by, say, 10%, then the true benefit of the smaller aircraft is no longer the 5% shown in the fan diagram, but more like 15%.

Equivalent-Cost Modelling

Having referred above to the problem of assessing the economics of aircraft based on cost alone, we now need to consider how to introduce revenues into the equation. Aircraft evaluation needs differ from those of airline route managers, for example, who focus on overall revenues and profits generated for their area of responsibility. Also, revenue management teams concentrate on the broader network picture in optimising revenue and capacity.

What the fleet planner needs to isolate is how much revenue is being generated purely as a result of a particular aircraft type decision. We need to identify the effects of frequency of operation and whether passengers exhibit a preference for a particular type of aircraft, even it is a broad preference for a wide-body over a single-aisle, for example. By the time we are analysing the economics of various aircraft types, we should already have conducted our market surveys and formed a good opinion of how aircraft size and frequency decisions can affect our growth, market share and profitability.

One useful way of integrating a revenue calculation into a standard cost comparison is with the concept of equivalent cost. We know that different classes of an aircraft generate different yields. Rather than divide the aircraft cabin into a

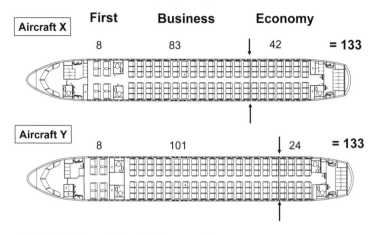

Figure 6.17   The Trap of Seat-Kilometre Costs

huge number of booking classes, as a revenue management system would do, we can simplify our approach and work on the basis of the physical classes of the aircraft and use average yields within each compartment. To illustrate the concept, we shall take two 133-seat configuration aircraft, as in Figure 6.17.

The layout at the top of the figure, Aircraft X, differs from the lower layout, Aircraft Y, in that the class divider between Business and Economy is further forward. Also, the galley and toilet positions are slightly different.

It can be argued that the cost of operating these two aircraft would be slightly different in that the Business-class product is higher than that of the Economy-class due to a different catering standard, for example. However, many economists would not disagree that these two aircraft would generate virtually identical unit costs. They would certainly appear that way on the traditional fan diagram.

However, what is missing is an appreciation of the revenue-generation *potential* of the two layouts. To better appreciate revenue potential we should choose one of the aircraft compartments, say the Business-class, and assign a coefficient of 1.0. If we believe that the average yield generated in First-class would be 80% higher than that of Business-class, and if we believe that the average Economy-class yield would be 60% of our datum, then we can reassess the earning ability of the aircraft, as seen in Figure 6.18.

The calculation suggests that Aircraft Y has an overall equivalent additional capacity of 7.2 Business-class seats. This is achieved simply by throwing a weighting towards the front of the aircraft to reflect the higher potential yield. Of course, this does not mean that we can necessarily *earn* the extra 7.2 equivalent Business-class seats. The answer to that question lies in the market situation, fare levels, strength of the competition and how revenue management specialists manage the overall capacity. What we have nevertheless achieved is a modification to the way in which two aircraft with an identical number of seats is measured.

The concept can be further developed by including underfloor space in the calculation, so that each potential tonne of cargo is converted into an equivalent number of Business-class seats.

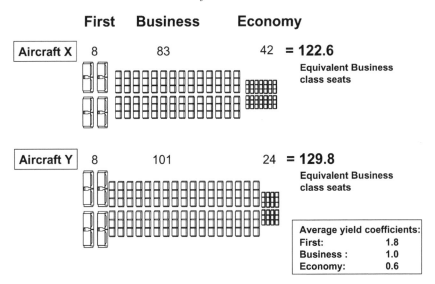

Figure 6.18 Equivalent Seat Revenue Potential

The fan diagram of comparative economics can then be modified according to the number of equivalent seats in each aircraft in the comparison.

In the above example we considered a comparison between two aircraft with the same number of seats, but with different earning potential. The concept of placing a value on seats takes on a new meaning when we compare aircraft of different sizes.

Markets do tend to grow at different rates, because they are driven by a variety of parameters. High-yielding premium markets are influenced by business activity and low-yielding leisure markets are influenced by disposable income, for example. Our market segmentation will assist in determining the rates of growth, which tend to be lower for the premium markets. The effect of this disparity between growth rates means that the proportions of First, Business and Economy classes might not be respected for larger aircraft. Thus, when comparing a 400-seater with a 300-seater, for example, we may wish to keep the numbers of premium seats fairly close, and fill up the extra space on the larger aircraft with proportionately more Economy-class seats.

The effect is that average potential yield will be lower for the larger aircraft. Any yield erosion is mitigated by two counterbalancing effects. Firstly, there are scale economies of operating a 400-seater of an equivalent technological standard to a 300-seater. Secondly, the planning load factor of the low-yield cabin will be higher, owing to the less dramatic effect of demand spill. Thus, it should be possible to achieve higher load factors with the larger aircraft.

Having said all of that, there is still the fundamental question of whether a jump in capacity from 300 to 400 seats is appropriate for the market and whether any sacrifice in frequencies would be necessary in order to moderate the capacity increase. This, in turn, could jeopardise market position in a competitive environment. Just because a larger aircraft has lower unit costs does not automatically make it more profitable, especially if significant differences emerge due to the higher yield traffic attracted to a daily compared to irregular flights.

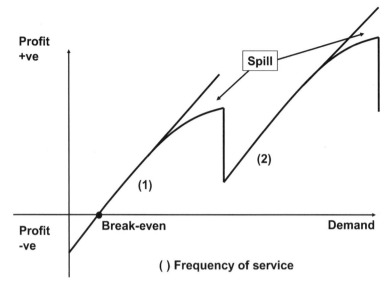

Figure 6.19 Proft Profiling (1)

Profit Profile Modelling

Another very useful modelling tool is the profit profile. Again, this is a way of taking revenues into account, with the added bonus of also incorporating a spill calculation.

The essence of the profit profile is to enable us to compare aircraft of different sizes and economic performance against original demand. As demand increases the profit profile suggests when it is economically better to either increase aircraft size or add additional frequencies.

In Figure 6.19 a single aircraft serves the market. When demand is zero the aircraft theoretically generates no revenue so the greatest '-ve profit' in the chart equates to the operating cost of the aircraft, however it is defined. As demand increases so do the profits, according to whatever yield function is defined. The most usual method is to assume an average yield for the whole aircraft, irrespective of demand level and market segment changes.

There comes a point along the demand curve when the average demand is such that spill is generated, according to a predefined coefficient of variation. This is where a curve will appear as the difference between average demand and actual traffic becomes apparent. At some point further along the demand curve, when it is judged that spill is becoming too great, a decision must be made concerning capacity. In the example, a second frequency is added at that point, thereby instantly reducing the profits to a lower level as the costs of operating the second aircraft now must be considered. As demand continues to grow, so does the profitability until, once again, spill begins to appear. The cycle can continue with more and more frequencies being added.

In Figure 6.20 we compare two aircraft of differing sizes. Owing to its smaller size, Aircraft A achieves break-even earlier than the larger Aircraft B. However, its smaller size means that Aircraft A starts to generate demand spill earlier. As demand

The better aircraft is.....

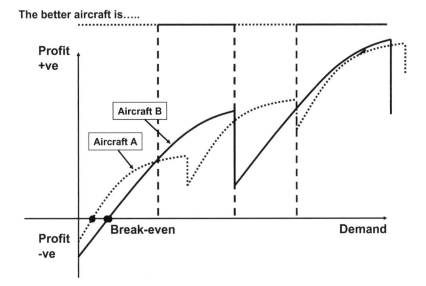

Figure 6.20   Proft Profiling (2)

grows it can be seen that different combinations of frequency generate the greater profit. Thus, profit profiling can be a good guide as to which size aircraft might be economically better according to expanding demand.

As always, there are some important *caveats*. Firstly, wide variations in average demand could easily mean that the market mix could completely change, invoking very different average yields. One way around this would be to modify the straight-line part of the profit curve to reflect gradual erosion of yield as a function of demand growth. This is termed 'yield bleed'. A second problem is that profit profiles do not take into account the demand stimulation effect of the provision of additional frequencies. Comparing a two-frequency operation with a single frequency ought to mean that different demand could be generated. Thirdly, no account is taken of network effects, whereby it might be essential to provide a certain degree of capacity on a route owing to the need to feed a hub, for example. Lastly, profit profiles should be created for both directions of a route, so there is no point in optimising for a single direction if that aircraft would be compromised on the return leg. We should resist the temptation to simply aggregate data because that would merely destroy our ability to model specific demand and revenue traits.

Sizing Models

Another useful stand-alone modelling technique concerns the assessment of different aircraft size against break-even loads. When contemplating growing aircraft capacity, we have to be careful that the larger aircraft does not suddenly make the route, or operation, unprofitable.

There is a golden rule in aircraft sizing which says that when a capacity shift takes place, the profit generated by the smaller aircraft should be at least equivalent to the

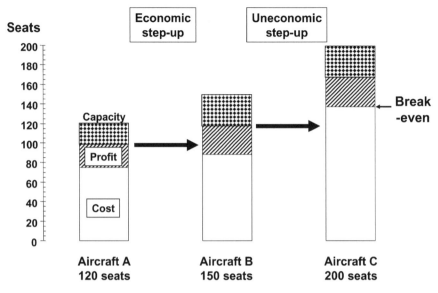

Figure 6.21   Economic Size Modelling

break-even point of the larger aircraft. In Figure 6.21 we can observe that a swap of capacity from Aircraft A to Aircraft B, undertaken at the moment when demand spill is too high, is economically viable. However, moving from Aircraft B to Aircraft C would not be wise as the operation would immediately become unprofitable. All the usual exemptions to this rule do apply, however, such as there may be overriding network reasons why a shift in capacity is necessary.

Aircraft sizing modelling has assumed new significance with the influx of multi-sized aircraft of virtually identical technology. Such aircraft have been designed so that their spacing *does* allow economic inter-changeability. Fleet planners would nevertheless do well to keep a close check on the actual configurations adopted in differently sized aircraft as well as calculating carefully the break-even load factors and their trends.

Dynamic Fleet Management

A new branch of aircraft economics burst upon the scene toward the end of the 1990s. The seeds of dynamic fleet management were first sown when manufacturers began designing operational commonality into their product lines. It is therefore somewhat surprising that this optimising technique, known by a variety of names, has taken so long to capture widespread acceptance. The technique has been variously called 'Demand Driven Dispatch' (D^3) by Boeing, 'Adaptive Aircraft Assignment' (A^3) by Sabre, and 'Dynamic Capacity Management' by Ortec of the Netherlands.

The concept involves making a continuous analysis of demand variations throughout the network and adapting the way in which each aircraft type is assigned to individual sectors.

Dynamic fleet management is an operational mechanism to better match supply and demand, not by influencing that demand through revenue management systems,

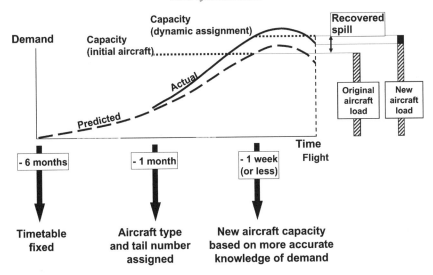

Figure 6.22 Single-Flight Demand Build-Up

but by manipulating capacity according to late changes in demand. So, once again, we return to our old friend demand spill, which underpins so much of aircraft economics and fleet planning. Our goal is to achieve an overall optimisation of profitability for the entire network through the best balance of supply and demand. This may mean accepting less than ideal results in one part of the system in order that the network may benefit.

There are two important pre-requisites for the concept to work. Firstly, a route network must offer aircraft swapping opportunities. A linear network, especially long-haul, does not lend itself to aircraft swaps to take account of demand variations. Secondly, the aircraft technology should be so close that a change of aircraft type has minimal effect on the airline operation. Different aircraft types with identical flight decks lend themselves very well to dynamic fleet management.

How Demand Evolves for a Single Flight

In order to appreciate the way in which dynamic fleet management works we need to examine how demand typically builds up for an individual flight.

We might expect that the timetable has been fixed six months before the day of departure. Hopefully, the revenue management system will track bookings as they occur and produce a prediction of the final load, building-in known no-show and last-minute go-show traffic. Overbooking limits will be set as part of this analysis. At some stage both the aircraft and tail number will have been assigned.

Imagine that the original booking prediction has underestimated demand, which begins to rise to such an extent that the actual load could exceed the capacity of the assigned aircraft. There would be some logic in swapping this aircraft with one of a higher capacity in order to ensure that the higher demand can be accommodated.

The same mechanism would work in the opposite sense, where real demand turns out to be *less* than that originally predicted. In this case, a smaller unit of capacity

Aircraft planned

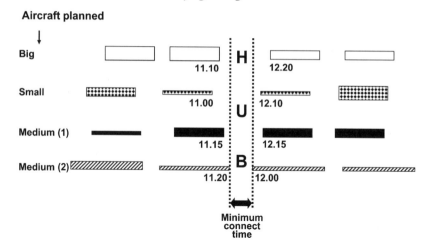

Height of bars represents demand magnitude

Figure 6.23 Looking for Swap Opportunities

would be assigned in order to achieve a higher load factor than would otherwise have been the case.

Swapping Aircraft on a Network Basis

Dynamic fleet management involves looking not only at a single flight, but at every flight in the network. Moreover, a solution found for one day might not be replicated the following day, especially if demand fluctuates a great deal.

On a multi-leg network, such as in Figure 6.23, we might have four lines of flight, operated by three differently-sized aircraft but sharing the same operational commonality, termed 'Big', 'Medium 1 and 2' and 'Small'. The seating differences between these aircraft should be small enough so that any substitution of the small aircraft by the large aircraft does not carry too big an economic risk. The original planning might have envisaged four flights coming into a hub, for passenger connections, and the four aircraft all continuing on other sectors.

In the diagram the thickness of the bars is a representation of demand for the various flight legs. What we can observe is that three of 'Medium 2's' legs are rather thin, and 'Small' is likely to face a capacity problem on its fourth leg.

Figure 6.24 shows the situation after swaps have been proposed. At first sight it may seem curious to propose that the largest aircraft takes over the line of flight of the smallest aircraft at the hub, when the onward leg is booked very low. However, the essence of dynamic fleet management is that the entire network has to be considered, not only for one day of operation, but for whatever period is appropriate to encompass demand variations.

In making a swapping recommendation, a dynamic fleet management optimisation tool would ideally need to examine demand on a strictly origin and destination basis, because a certain proportion of traffic flows into a hub with the sole purpose of

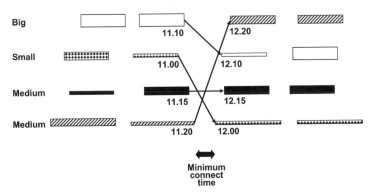

Height of bars represents demand magnitude

Figure 6.24 Making the Swaps

connecting into other flights. There is a correlation between traffic flows into and out of the hub. Indeed, the latest fleet assignment tools and revenue management systems also calculate on an origin and destination basis, so it is logical for dynamic fleet management systems to orientate themselves in the same direction. However, only a relatively small number of major airlines have the in-house capability to manage their data at this rather complex level. Having said that, current dynamic fleet management tools also calculate successfully on a flight leg basis.

The decision to swap is not just made on the basis of booking level, cost of operation and potential revenue. It is important to reflect in the calculation the revenue that would otherwise have been lost through spilled passengers. This is termed 'shadow cost' because it is not actually incurred, but merely influences the assignment decision. It is generally presumed to be equivalent to the revenue.

Potential Challenges in Dynamic Fleet Management

Making a change in aircraft assignment very close to the time of departure will have an impact on many aspects of the operation. The key to a successful assignment system is to ensure that any disbenefits are more than outweighed by the better matching of supply and demand that results.

One immediate challenge is that forecast locations of aircraft throughout the system will alter, complicating maintenance planning. This could be less of a problem for short-haul operations where aircraft would channel through the main base with more regularity than long-haul aircraft. Another problem is that a sufficient number of cabin crew would need to be maintained at swapping points throughout the network to ensure that substitution by the largest aircraft in the fleet could indeed be handled. Also, dispatch of cargo could be compromised if underfloor space suddenly evaporates due to a substitution of a smaller aircraft.

Sometimes, mundane issues such as having sufficient galley trolleys or baggage containers can compromise the desire to make a reassignment.

Another significant challenge is that the closer you wait to the departure time to make a swap, the more likely it will be that the booking level of the flight will exceed the capacity of the smallest aircraft in the fleet. This means that the sudden removal of that capacity could create over-sales. This creates a significant practical problem in that not all passengers who might have been given confirmed reservations can access the flight that is now served by the smaller capacity unit. There is no point in telling them that the aircraft have been changed to boost the profitability of the airline they have chosen to fly with! This is indeed the most serious objection to the use of dynamic fleet management in an assignment, rather than fleet planning, context. The only way around this problem is to only propose aircraft swaps up to the point where booking levels reach the capacity of the smallest aircraft in the fleet.

Continental Airlines perform swaps within their 737 types at up to 60 days before departure. As they use the same number of cabin crew on the aircraft, one of the potential hazards is not present. However, providing the correct catering levels is more of an issue. The ability to swap aircraft can be quite dependent upon the characteristics of the market. For example, Albuquerque and Mexico are candidates for capacity swaps because both markets tend to book a long time in advance, both are fairly volatile and neither tend to book at the same time.

Air France has the potential to introduce swaps with their A320 family aircraft but they have two issues to address. Firstly, their differently sized aircraft are operated with different numbers of cabin crew, and secondly, there are union agreements that preclude cabin crew changing planned destinations.

Finally, constantly changing capacity makes it more difficult to compile a demand forecast. Forecasting becomes so much easier if the demand is as unconstrained as possible, yet constant tampering with capacity is always altering the final loads carried. However, one could level the same criticism at revenue management techniques.

Perhaps one of the biggest obstacles to the development of the technique is airline organisation. Silo-orientated organisations are unlikely to embrace the technique as cross-functional cooperation between, for example, departure control and revenue management groups, is vital.

Several North American and European airlines have practiced a form of dynamic fleet management, including Continental Airlines, American Eagle, Swiss, KLM and Finnair.

Using Dynamic Fleet Management in Fleet Planning

Dynamic fleet management techniques can also be successfully applied as an aid to fleet planning. The main difference between these approaches concerns the origin of data and their treatment, and how close to the final day of departure a swap decision can be simulated.

The differences in data concern, for example, the calculation of yields and demand data. In the operational context, real yields from the revenue management system can be used, whereas in a planning context a generic yield linked to sector length will suffice. Operationally, real demand should be extracted from the reservations system, whereas for fleet planning purposes we must resort to average demand levels, with a coefficient of variation to explain deviations from the mean. The most usual way of creating demand data for a study period would be to use a Monte Carlo simulator to produce a quantity of randomly generated values, or trials. These can then be compared with aircraft capacity on a daily basis to evaluate how to achieve a better match.

Some More Drivers of Aircraft Operating Economics

We have already seen how airline network choice and the operating environment combine with aircraft choice to give an accurate picture of economics. During the discussion on individual cost items we addressed some of the forces which shape the magnitude of the DOC breakdown. To finalise this review we shall explore some further factors that may also affect aircraft selection and profitability. Many of these will become more and more significant over time as it becomes increasingly challenging for manufacturers to offer large-scale economic improvements by technological breakthroughs.

Scheduled vs. Charter vs. Low-Cost Operation

Much of our discussion has centred on the scheduled business. However, we must recognise that the leisure business operates to a different set of priorities. Scheduled airlines organise their entire planning process around the business cycle with the inevitable result that the two become uncomfortably linked. A feature of the leisure business is that it is robust, stable, growing and not necessarily linked to the business cycle.

We have seen that utilisation is an important parameter in spreading fixed costs. Short-haul scheduled operators can sometimes struggle to reach a utilisation of even five hours/day per aircraft, according to their market needs. However, leisure airlines can regularly achieve up to 17 hours per day in the peak period from their aircraft. This unique feature, coupled with average load factors of around 95% and dense seating configurations, means that the economic picture is rather different. Such high utilisation can only be achieved with extraordinary levels of reliability, which themselves are correlated with young fleets. So the prioritisation of key buying factors for a charter airline can be expected to be somewhat different.

The low-cost carriers have emulated many of the principles that have made leisure airlines so successful. By reducing passenger amenities, and coupling high utilisation and high frequencies with a low-fare strategy, these airlines have challenged traditional operators to such a degree that the airline economics landscape has changed. Key to their success have been their fleet selection policies which, like those of the charter operators, have concentrated on aircraft types with high reliability. One school of thought that took hold in the early days of low-cost carriers was that only a single-type fleet would work effectively. This is certainly not the case. Southwest Airlines operates a single-type fleet due to historical reasons. As Southwest were developing their fleet, they had no realistic alternative to the 737. Now it is simply too late to change. However, the newer breed of low-cost carrier has had the luxury of a real choice of aircraft type.

There are two basic reasons why low-cost carriers may operate more than one aircraft type. The first reason concerns the value of contestability. When easyJet were expanding their operation they could have simply expanded their fleet of 737s. However, Airbus entered the competition with the A319 and the airline found that any potential disbenefits of operating two types of aircraft to serve similar markets would be offset by the lifetime value of the effect of having the two manufacturers compete vigorously for the business. In any case, as the overall number of aircraft expands, the *incremental* effect and value of commonality diminishes.

The second reason why low-cost carriers consider a second aircraft type is where a secondary market may be developed. JetBlue Airways had been operating a fleet of A320 aircraft quite successfully before adopting Embraer 190s in order to develop a market that would have been too small for the larger A320.

The Effect of Fleet Size on Economics

A small airline serving a small market has limited opportunities for creating the same synergies as a large airline. The existence of economies of scale gives large established airlines a head start in economics over small competitors. Some of that advantage can be eroded because smaller businesses should be more nimble in decision-making and carry less overhead.

The real issue in fleet planning is one of creating a fleet with as much diversity as possible to serve a broad spectrum of markets, but retaining as much similarity as possible from a technical and operational point of view. Again, we return to our old chestnut of commonality. The economic benefits of fleet operational commonality can never be understated. These benefits affect, for example, the numbers of crew required to service a multi-size fleet, crew training, spares investment, ground service equipment and handling requirements.

Coupled with the advantages of commonality is that start-up costs become a huge area for potential investment saving. It is wise to analyse the manufacturers' abilities to rapidly supply spare parts. A review of overall customer support, including spares pricing policy, ordering mechanisms and delivery record is important. Access to a spares pool managed by other operators of similar aircraft types should be a part of the overall picture, too.

A useful way of assessing the overall impact of fleet size on economics is to integrate the various factors into a single explanatory curve. Thus, we could consider spares investment, maintenance shop investment, the number of flight crews and their training, the number of simulators, and the number of maintenance crews and their training, for example. Figure 6.25 illustrates the principle of how economies accumulate as a function of fleet size.

The Impact of Alliances on Economics

The history of airline alliances confirms that one of the most essential strategic ingredients is an escape path in case something goes wrong with the relationship, or else the strategy of one of the partners alters.

Despite what appeared to be long-term commitments on the part of both KLM and Alitalia to co-operate technically, their technical cooperation alliance in the late 1990s fell on stony ground very rapidly. Aer Lingus withdrew from the Oneworld

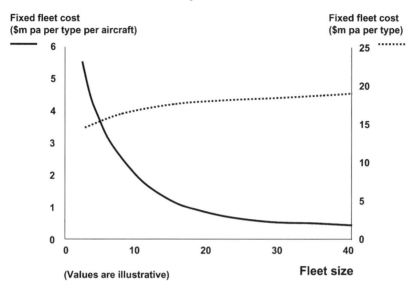

Figure 6.25 Fleet Size Economies

alliance due to a major shift in their strategic direction. Although it was not the case, it could have been disastrous if Aer Lingus had allowed their fleet strategy to be driven by their alliance partners, only to find themselves compromised upon leaving the alliance.

Airlines undoubtedly reap revenue benefits due to alliance membership but true cost savings in big-ticket items such as fleet acquisition may be out of reach for all but a handful of airlines.

There will always be institutional obstacles, such as union issues and manufacturers' pricing strategies, to airlines reaping the fleet cost savings that might be apparent on paper.

The Problem of External Factors

Airline economics have always been affected by external factors, some of which are conditions of the geographical area in which airlines operate and some of which just conspire to make life difficult. The huge fuel price increases of the 2000s affected the entire industry, and even those airlines able to hedge the majority of their fuel bill were simply postponing the moment when the higher fuel prices ate into profit margins.

An underestimated external effect on economics is the effect of exchange rates. Airlines usually incur a high proportion of costs either in US dollars, such as loan or lease payments, or else the currencies of the countries in which they do business. Revenues, however, tend to be biased toward the local currency of the airline. It is sometimes the case that an airline's profitability is completely overwhelmed by exchange rate losses. Indeed, agonising over 1 or 2% in DOC differences between two aircraft types pales to insignificance against external effects.

Some Solutions to Improve Operating Economics

Outsourcing of certain airline activities has become a popular method of attacking cost levels. Non-operating parts of the business, such as revenue accounting, have become obvious targets. However, from the aircraft perspective, it is sometimes more profitable to break from the traditional notion of owning and managing every activity.

Areas ripe for outsourcing include training and maintenance, where many opportunities exist to contract complete support packages at market-driven rates. Division of the airline business into separate business units has proved to be a popular way to engender efficiency.

Airlines serving a multitude of market segments have the ability of switching resources if a downturn appears in one of them. The Asian economic crisis of 1997 held purely Asian carriers in its grip, whereas international carriers redirected their aircraft to more profitable markets. Qantas redeployed 747s to South America to await the Asian recovery.

Diversification, whilst a useful method of spreading risk, can sometimes be extreme. One former private airline in India had interests in windmills, water bottling and a television station! Kingfisher Airlines is a successful Indian airline, but the owner's initial (and continuing) focus has been on beer.

Some assessment of risk should be included in an economic analysis. For example, if market conditions worsen, we do need to have a good appreciation of the fixed and variable cost relationship. Older aircraft, which may be fully depreciated and paid for, can sustain inactivity more easily than new aircraft. Yet, this advantage must be balanced against the risk that fuel price increases will have a greater effect on older aircraft than newer aircraft.

Spares Provisioning

Much of aircraft economic analysis is concerned with the understanding of actual operating costs. However, the spares provisioning is of vital importance and should be included in the overall cash flow analysis.

Specialists will determine the spares provisioning for a fleet by reference to the ATA numbering system. A *poisson* formula is applied to calculate the probability of failure of any component during repair time. Through this, a specialist can determine the protection level. Where, for example, the protection level is 95%, we can be confident that we are at least 95% certain of having enough parts in stock to survive the repair time of any failed part.

There are some fundamental difficulties with this approach. Reliabilities vary dramatically between airlines owing to differing maintenance practice and modification status of the aircraft. The reporting of reliability data is often incomplete and limited to 'serialised' parts, making it very difficult for spares provisioning specialists to recommend parts holding.

Furthermore, the protection level principle sometimes suggests the purchase of expensive items just to achieve the 95% hurdle. If an airline is already achieving, say, 94% it may be more prudent to disproportionately improve the protection on cheaper parts and get a better protection for less investment.

In fleet planning, it is more usual to use 'rule of thumb' than the rigorous system used by the specialists for actual fleets. This means taking a proportion of the fully-equipped aircraft price for airframe spares. This value has historically ranged from 8% to 10%. Engine spares and spare engines might add another 5–7%. Over time, initial provisioning investment levels are reducing owing to better reliability, shorter lead-times for delivery of spares and, indeed, the desire to hold fewer components in stock.

It is not necessarily the case that an airline wants to invest in spares. Pooling, third-party suppliers and leasing from the manufacturers are alternatives to consider. Pooling might be a solution for the operator of an aircraft which serves thin niche markets spread over a wide geographical area.

The manufacturer offers a variety of spares services to its customers. For example, a growing number of airlines have elected to have their spares supplied directly, and on a 'just-in-time' basis from Airbus. The manufacturer assumes responsibility for administration and tracking of shipments and also benefits from economies of scale. These advantages are passed along to the customer, who incurs less overall cost.

Boeing offers an all-encompassing maintenance and materials management service for the 787, which extends the engine manufacturers' power-by-the-hour innovations to the airframe. The manufacturer works closely with key suppliers to the aircraft programme, and is able to help customers with predictable costs, whilst managing the actual transactions.

Another formula is for airlines to acquire spares from a locally-based consignment stock on a pay-as-you-go basis.

In Summary

In this Chapter we have seen the importance of making accurate assessments of areas such as interior configuration and aircraft performance. Economists have a habit of disagreeing when it comes to defining a classification of aircraft economics, but we must navigate the hazards and establish a customised cost breakdown appropriate for our specific circumstances. We have seen how it is useful to get a feel for what *drives* economics of aircraft. Before blaming the aircraft, we need to understand the effect that the operating environment and route structure may have on results.

Economic analysis is never complete without an assessment of revenues, and there are various useful stand-alone modules, such as profit profiling, which can assist in comparative studies.

New techniques in fleet planning and aircraft assignment, such as dynamic fleet management, are expanding the scope of economics and helping us better appreciate the flexibility built into aircraft families.

Lastly, we need to appreciate the effect of fleet size on overall economics, together with the impact of spares provisioning. Although aircraft economic analysis is a great help in positioning different products, it is not enough to help us make decisions for the longer term. For this we need to perform an investment appraisal and bring in the overall life cycle costs of the investment. We will now turn to this final stage of the evaluation process.

Chapter 7

The Investment Appraisal

Drawing the Threads Together

Having spent a lot of time looking at snapshots of operating costs, it's time to take in the movie and look at the impact of an aircraft operation over a period of many years.

A fundamental precept of fleet planning is that it should be continuous. Naturally, we may need to undertake specific and in-depth analyses and comparisons of aircraft from time to time to implement planned rollovers of the fleet. However, it is important to maintain a continuous watching brief on new technology developments from the manufacturers, in case there are opportunities to introduce improved versions of their products at attractive prices.

Also, we should be on the lookout for opportunities to improve the financing structures and reduce financial exposure. We may wish to consider sale and leaseback dealing, reconfiguration of the fleet to support a rebranding exercise, or simply adding another unit to the fleet to accommodate natural growth. All of these actions will have an impact on the balance sheet over a long period of time.

Indeed, *any* potential investment should be evaluated over a period of time. In order to correctly assess the impact of a fleet change on the business we will need to broaden our view of aircraft economics beyond purely operational issues.

This Chapter will address the scope of an investment appraisal concerning a change in an airline's fleet, and examine the most common methodologies applied in order to value an investment over time. We shall see that the investment appraisal alone cannot suggest the best solution for the fleet, but without it we would be unable to recommend the optimum investment from a cash flow perspective.

The Impact of Aircraft Price in Aircraft Selection

Ownership charges, comprising depreciation, interest and lease rentals, account for 12.5% of total operating costs according to IATA in 2004. The burden is accentuated when we add interest charges on loans. Depreciation is a function of the price paid for the asset, of course, and it is worth considering the elements that go into aircraft price and the extent to which they can be influenced.

Order Size

In the same way as we can expect a discount for purchasing in bulk in our supermarket, aircraft are also priced attractively when bought in bulk. Part of the logic is that the manufacturer's customisation costs for a deal involving a single aircraft are not far away from those involving a large order. Also, it is comforting for the manufacturer to book a large number of units as it secures jobs and engenders confidence in the company. Another important factor in pricing concerns the balance between firmly ordered and optioned aircraft, along with any buy-back or residual value guarantee conditions, for example.

Concessions

These may be offered as part of the usual battle between competing manufacturers to win strategic orders, and are often available to encourage customers to place orders to launch new models. Clearly, the potential and lifetime strategic value of a customer plays a significant role here.

Price escalation

Owing to the time difference between committing to a purchase and actually paying money, it is important for prices to be escalated, and these must be applied in the investment appraisal.

Changes to Specification and Buyer Furnished Equipment

Here is an area where the airline can control its investment. Enrichment of the basic aircraft is considered highly desirable to gain a market edge, but it can still add millions to the price. The residual value of the aircraft can also be affected according to the degree of customisation.

Support Package

A new customer can expect to negotiate a support package involving on-site assistance and training for a defined period.

Scope of an Investment Appraisal

Everything that we have examined in earlier Chapters has concerned the air transportation part of the business. We have looked almost exclusively at costs and revenues pertaining to the operation of a particular aircraft type. Sometimes we have strayed into indirect operating costs, because we often need to have a vision of the magnitude of aircraft-related costs to total operating costs. Now, we need to broaden our vision even more and look at the non-operating side of the business, especially where aircraft choice has a role.

Non-Operating Activities

Maintenance contract work   Maintenance expertise will be built around the type of aircraft operated by the airline, and it is logical to use this expertise to offer third-party contract work to operators of similar aircraft. This widespread practice ensures that workshop facilities are fully utilised, for example, and should be a useful source of revenue. One could also argue that an airline responsible for the maintenance activities of other operators is able to extend its influence in the region to a certain extent.

Simulators and crew training As with maintenance contracts, an airline with knowledge of a particular type of aircraft could sell simulator time to other operators and engage in more general crew training.

Leasing activities If demand fluctuates a great deal throughout the year there may be opportunities to lease-out a part of the fleet during the low periods to airlines needing a particular type of aircraft.

Trading Sometimes, significant non-operating revenues can be gained simply through the trading of aircraft in the fleet.

All of the above non-operating areas are indirectly affected by the choice of aircraft. Some of them, such as trading, are not really predictable, but many other activities can be brought into the calculations.

Segregating Profit and Cash

The main tool of the economic analysis is the profit statement, which helps measure the efficiency of the operation. The main tool of the financial analysis is the cash flow, which helps measure our financing, cash and equity requirements. The distinction between profit and cash is mostly driven by tax rules and timing, as airlines can be cash-rich or cash-weak depending upon the balance of money owed or paid. Highly seasonal charter airlines tend to be victims of this pattern. It is a quirk of airline economics that the business can seemingly be unprofitable for long periods, yet still generate a great deal of short-term cash.

The Profit Statement In order to arrive at the after-tax profit, we should segregate the operating and non-operating profit, which may be subject to different tax rates. Also within this segregation we should identify both cash and non-cash items, as seen in Figure 7.1.

The Cash Flow There are many different types of cash flow, serving different purposes. The operating cash flow includes cash operating costs and revenues associated with the core business of transporting passengers and, if appropriate, cargo and mail. The investing cash flow involves that money associated with the purchase and sale of assets, including not only the aircraft but also any investment in equipment or property. It does not tell us anything about the prosperity of the business. The financing cash flow concerns money from lenders and investors in the business. Inflows of money are in the form of loans and outflows would be repayments. Another flow of cash to be considered is that connected with the payment of taxes. Keeping all of these cash flows separated is extremely important.

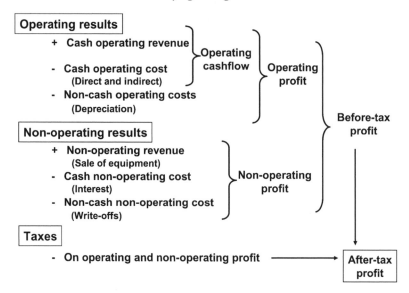

Figure 7.1   The Profit Statement

Elements to be Included in the Investment Appraisal

There is a tendency to think of the aircraft and its related investments, such as spares, as the only item to be considered in the appraisal. The investment project may comprise a whole series of subsequent and related investments. These might include the future acquisition of spare engines as fleet size grows, simulators, hangars, and type-specific tooling.

It is not necessarily the case that an investment appraisal should include revenues. Quantifying additional revenues as a result of installing seat-back videos or even a more general cabin reconfiguration, is very difficult.

We should also decide whether we are going to project our costs and revenues into the future with inflation considered or not. There is no right or wrong here. However, if we intend to use current rather than constant prices, then a forecast of inflation rates is essential. In the same vein it might also be essential to forecast exchange rates with other currencies with which we might deal. In addition, we will need to include future tax payments on profits, which is another tricky area.

Fundamentally, we will need to forecast all of our costs and revenues over a defined analysis period. In Chapter 3 we considered how our fleet plan might look as a function of forecast demand growth. This can certainly form the basis of our leap into the future. Now we need to expand our economic analysis by forecasting parameters such as fuel price, labour rates, maintenance material costs, landing and navigation fees, catering costs and so on. The yield forecast should be linked to our future competitive position as well as exchange rates, GDP and disposable income evolution. There is no doubt that the scope for getting things wrong is huge.

Conducting an investment appraisal enables us to embrace changes to the magnitude of economic parameters that cannot be seen in a simple snapshot. Thus,

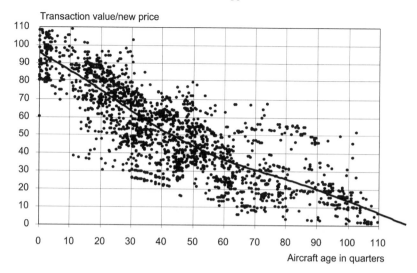

Figure 7.2   History of Narrow-Body Trades
Source: PKAirFinance

we must now take into consideration the total life cycle cost of our investment. We saw in Chapter 6 that maintenance costs are usually lower than mature levels in the first few years after a new aircraft enters the fleet due to the combined effect of warranties, the learning curve and newness factors. This honeymoon period can be calculated for a particular operation as a series of coefficients to be applied to the annual mature costs for the first few years of operation. Similarly, at the other end of the scale, we have the opportunity to reflect higher maintenance costs, and even fuel burn, as the aircraft gets towards the end of its economic life.

In Chapter 6 we also concluded that the overall size of a fleet affects the total economic picture. Again, it is in the investment appraisal that we can encompass all of the ancillary investments that go hand-in-hand with the aircraft. The initial provisioning of spares for a large fleet size will be of a lesser order of magnitude than for a fleet made up of diverse types.

Another element which needs to be incorporated into the investment appraisal is the residual value remaining in the investment at the end of the appraisal period. Residual value is sometimes very high on a buyer's list of key decision criteria. This is especially true of leasing companies, which view an aircraft more as an investment than a tool to generate traffic.

Residual Values

Some Definitions

Value means different things to different people. The accountant will think of it in terms of 'book value,' or the value recorded in the accounts. An aircraft trader will consider it as the 'fair market value' of the asset under prevailing market conditions.

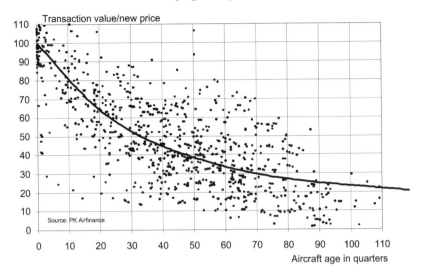

Figure 7.3   History of Wide-Body Trades
Source: PKAirFinance

The professional organisation giving advice on aircraft trading is the International Society of Transport Aircraft Trading (ISTAT). They define 'base value' as being the appraiser's opinion of the underlying value of an aircraft in an open, unrestricted, stable market environment with a reasonable balance of supply and demand, and assuming full consideration of its 'highest and best use'. ISTAT go on to say that an aircraft's base value is founded in historical trends and in their projection, and presumes an arm's length, cash transaction between willing, able and knowledgeable parties, acting prudently and with an absence of duress.

It is worth noting that, regrettably, many transactions do take place with some degree of duress.

Figures 7.2 and 7.3, compiled by Nils Hallerstrom and Jan Melgaard of PK AirFinance, reveal the wide disparity in traded values of narrow-body and wide-body aircraft as a function of age. Clearly, we need to build an understanding of the reasons for the dramatic scatter of points on the diagrams.

Factors Influencing Residual Value

Age   The age of an aircraft can explain anything from 50% or more of the value of an aircraft. However, age is far from being the most reliable indicator, especially as residual values tend to behave in a cyclical pattern. The sometimes wide disparity between transactions for similarly aged aircraft can be explained by differences in modification status, maintenance levels and the number of aircraft involved. Aircraft age alone is not enough to determine an aircraft value.

Production line position   Early aircraft of a new type off the production line are sometimes infected with the 'dash one-hundred virus'. This is where a small number of aircraft have been produced with lower design weights, which are often superseded fairly quickly when manufacturers use up their engineering margins and rapidly improve the systems status. There is nothing inherently wrong with any aircraft with a 'dash one-hundred' suffix. However, they do carry a stigma in the eyes of some and residual values may be affected.

Production status   Of far more importance than 'dash one-hundred virus' is the status of the production line itself. As soon as an aircraft production line closes, perhaps because a newer model has been launched into production, then residual values can be expected to suffer. It is therefore always a good idea to check the ordering patterns and backlog of an aircraft variant carefully to ensure that it really will be around for a long time to come.

   Also, the very last aircraft off the production line cannot be expected to hold their initial values for a long time.

Inflation   Trends in residual values are more apparent with inflation removed from the picture. It is real economic conditions that will determine whether an aircraft is likely to be attractive. Also, keeping inflation in the residual value forecast means that the inflation values themselves must be forecast.

The growth of the economy   Unsurprisingly, the demand for new aircraft tends to rise during an economic boom whereas it is second-hand aircraft that are in demand during a recession. Despite this generality, there are the usual regional differences to consider. For example, during the Asian crisis in the late 1990s the market for single-aisle aircraft was not significantly affected because the Asia market is made up largely of wide-bodied aircraft.

Prices of new aircraft   There is a tendency for new aircraft prices to set the ceiling on market values, especially if prices are stable. The effect could be that very new aircraft purchased for relatively higher prices might find their values quickly eroded.

Interest rates   When interest rates are high the pricing of second-hand aircraft is likely to rise as well to help recover the higher costs of financing, or leasing, incurred.

Aircraft economic performance   Efficient aircraft are clearly more likely to have better values than less efficient aircraft. Sometimes the economic attributes of an aircraft can be drawn out by an extraneous event. For example, high fuel prices enhance the value of more fuel-efficient aircraft.

Condition of the aircraft   Just as we expect to pay more for a 'low mileage, one careful owner' car, we should not be surprised to see aircraft which are well-maintained and with low flight hours and cycles command a premium. Other attributes that enhance value would be the incorporation of all relevant airworthiness directives and service bulletins, and an impeccable maintenance history. Also, aircraft which have spent their lives operating in stable environments tend to retain their value more easily.

Commonality   Aircraft belonging to a family with similar technology hold their value far better than an orphan, which may be more difficult to adopt.

Flexibility   This is linked to some degree with commonality. Aircraft which can be more easily deployed in alternative regions and markets are more attractive because they can be found a new use more easily if conditions change. The larger the aircraft, the more difficult it can be to locate a new market with the right level of traffic.

Major fleet re-equipment policies   Should a large carrier implement a major re-equipment programme, this could result in the market being flooded with a particular type which will then be saddled with depressed underlying values.

Stability of the manufacturer   The more likely the manufacturer will stay in business, the more likely that aircraft values will remain stable. The issue concerns long term support of the product. With the disappearance of the manufacturer, support may fall into the hands of an organisation that has other priorities than to support an inherited product.

Depreciation, Loan Payments and the Base Value

An aircraft's base value and depreciated book value can be quite far apart. Most of the aviation industry has adopted the simple straight-line method of depreciation. Thus, if we were to depreciate the aircraft down to zero in a 20-year period this would mean an annual depreciation charge of 5% *per annum*. This suggests that the aircraft would be worth precisely 50% of its original cost after 10 years.

However, base value curves may place the aircraft value at any point above or below the straight line at any point in time. Also, the market value of the aircraft may be either below or above the repayments owed on the aircraft, depending upon the type of repayment scheme being applied. Figure 7.4 shows stylised aircraft value behaviour and where potential areas of risk may emerge according to changes in the market value over time.

In the first few years it may be the case that the value of a new aircraft declines, even if market conditions do not vary. This is because a new aircraft may have been subject to a costly customisation, and it would not be worth the expense to rapidly change the configuration to suit another operator. Thus, new owners would expect a significant discount to be offered to compensate for another customisation. Once the discount value has been reached, the price should start to rise and, depending on the rate of inflation, might continue in an upward direction for many years. Thereafter there might be a period of stability as the costs of ageing are being balanced by the rising price of new aircraft. Beyond an age of around 15 years the value will start to decline as retirement is anticipated. The most unstable period is towards the end of the life of the aircraft when the value can drop very suddenly due to, for example, new noise legislation or a sudden rise in fuel price.

However, at any point in the life of the aircraft the value might be affected by an economic recession, or the introduction of newer technology, which might even render the aircraft obsolete.

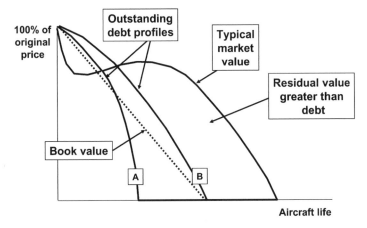

Figure 7.4   Value Behaviour – Example

Hopefully, the market value of the aircraft will be positive long after the asset has been fully depreciated. It can also be useful to plot various debt repayment profiles, such as 'A' and 'B' in Figure 7.4, in order to spot potential risk areas at different points in the aircraft life.

Why Residual Value is Important in Fleet Planning

Comparison of residual values between competing aircraft types is just as important as comparing their operating economics. Indeed, now that we are embarking on an investment appraisal over the life of the aircraft, it is essential. A good knowledge of residual value can help an airline determine the optimum time to introduce an aircraft into the fleet and can help in the construction of a financing package.

Also, selecting an aircraft with good residual value potential is a good investment and brings a degree of confidence into the planning picture.

The Life of a Project

Investment projects, like cats, have many lives.

Economic life   This is the period during which the project is deemed to generate an economic return.

Physical life   The structure of an aircraft can, in theory, always be made to be airworthy, but beyond a certain point the investment required to extend the physical life is no longer justified. The aircraft can then be said to be at the end of its physical life.

Technological life   Standards of technology can completely change within the lifespan of an aircraft. The emergence of fly-by-wire technology, with its inherent benefits in terms of reliability and maintainability, has meant that traditional flight control systems are viewed as a very different level of technology. Sometimes, as in the case of the development of a two-crew cockpit, a technology development can render aircraft virtually obsolete.

Depreciation life   The period over which an aircraft is depreciated is not at all concerned with technology, but is linked to some degree with economics. As depreciation is a fixed cost, the hourly operating cost can be reduced either through higher utilisation or else extending the depreciation period. At the other end of the scale there may be tax advantages in accelerating depreciation, which would reduce the operating profit but allow the owner to sell the aircraft at perhaps a substantial book profit and reinvest in new equipment sooner than might otherwise have been possible.

Financing life   This is the period over which any loans should be repaid. It is advisable that the investment appraisal period should at least be equivalent to the financing period.

Choosing the Right Decision Tool

A variety of methods of investment appraisal exist. These can be divided into the traditional methods, including payback and return on investment, and methods such as net present value (NPV) and internal rate of return (IRR) which both take account of the time value of money.

Payback

The payback method is whereby the accumulated forecast profits of an investment are compared to the magnitude of the original investment. The idea is to determine how much time may pass before the profits equate to the investment.

Although it is certainly useful to know that an investment can be recovered quickly, this method cannot predict what might happen to profits beyond the payback period and cannot tell us anything about how the value of money changes over time. Indeed, it is possible to rank projects incorrectly with the payback method where, for example, a project that pays back quickly cannot be relied upon to sustain profits in the longer term.

Return on Investment

This method also relates forecast profits to the original investment, but compares the average annual profit generated as a percentage of the investment.

A major problem with return on investment is that it is not possible to distinguish between large and small investments. A high rate of return could be achieved either way. For example, if you invest $4,000 that generates an average profit of $1,000 per year over five years, this gives us a return of 25% *per annum*. However, investing $40 million which generates $10 million per year over five years gives us exactly the same return on our investment. No account is taken of the amount of capital

$$NPV = \sum_{k=1}^{k=n} \frac{CF_k}{(1+r)^k} - I$$

CF_k = Cashflow at year k
n   = Number of study years
I   = Initial investment
r   = Discount rate

If NPV > 0 : The investment is worth doing
if NPV < 0 : The investment is not worth it

Figure 7.5   Net Present Value

employed and, as with the payback method, no account is taken of the time value of money.

The Time Value of Money

The most widely applied investment appraisal method involves net present value (NPV) or internal rate of return (IRR). We shall now examine the derivation of both of these decision tools.

Net Present Value

Discounting is basically the reciprocal of compounding. Thus, $1 is rated more highly today than at some point in the future. This is useful because we can reflect the erosion of purchasing power through inflation and uncertainty regarding the future. Moreover, NPV allows us to compare future cash flows on a common basis.

Figure 7.5 shows the formula for calculating NPV. The net present value is the summation of all future cash flows in the study-period, minus the initial investment. Our aim is to achieve a positive NPV for the project under study, or else choose the option that generates the highest NPV.

We can represent the cash flow pictorially by comparing our investment in the project at Year 0 to all the future cash flows, both in future value and present value terms. Figure 7.6 shows a simplified cash flow structure, where the first income is generated in the period after the investment has been made.

Selecting a discount rate   A vital component to NPV is the selection of the rate at which future money will be discounted. As the discount rate rises, future values diminish in value. Thus, the present value of $1 in five years' time with an 8% discount rate is valued at $0.68 today (see Figure 7.7). Yet a discount rate of 10% would mean that our $1 would be valued in five years' time at $0.62. So, high discount rates make investments harder to justify.

The discount rate is intended to reflect an alternative use for capital. Whatever we do with capital involves some risk and we must take account of this in our valuation

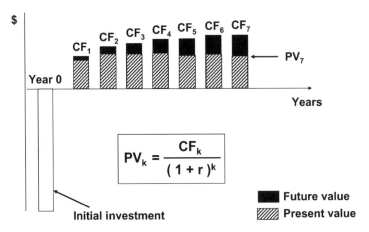

Figure 7.6   Cash Flow Structure

of a project. We could imagine plenty of alternative uses for capital that may generate high returns, at varying risk. The fleet planner is advised against suggesting to management that the capital to invest in new aircraft should be bet on a horse. However, one could argue that, as the opportunity rate of capital, the discount rate should be at least equivalent to the interest rate that one could have applied, were the money to be invested in the bank.

As big projects like aircraft investment carry with them some risk, the discount rate should be raised in order to reflect this. As we have just seen, raising the discount rate means that future money is valued even less, making it more difficult for a project to earn money in present day terms. The rate fixed may then be considered as a 'hurdle' rate which, unless achieved by the project, should not go ahead.

Although it is possible to fix a discount rate in an arbitrary way, there is a more scientific approach.

The WACC method of selecting discount rate   The WACC, or the weighted average cost of capital, is a way of computing a discount rate by blending the company borrowing rate with the shareholders' expected return on investment. We should now examine the components of the formula, which is shown in Figure 7.8.

R_d could be based, for example, upon LIBOR (the London Inter-Bank Offered Rate) plus a margin of (say) 2%. It is important to decide whether the company borrowing rate is being considered before or after tax has been deducted. If a tax rate of 35% is

$1 in five years' time discounted at 8%:   Present value $= \dfrac{1}{(1 + 0.08)^5} = 0.68$

$1 in five years' time discounted at 10%:   Present value $= \dfrac{1}{(1 + 0.01)^5} = 0.62$

Figure 7.7   The Principle of Discounting

$$\text{WACC} = R_d \times \frac{D}{(D+E)} + R_e \times \frac{E}{(D+E)}$$

R_d = Company borrowing rate
R_e = Shareholders' expected return on equity
D = Debt
E = Equity

Figure 7.8   The Weighted Average Cost of Capital

being excluded from the borrowing rate, this will give an advantage to the project. A contrary position would be to argue that tax is not an issue that should contaminate management's view of a project and that the pre-tax value should be used.

R_e is far harder to estimate. Shareholders' expected return comprises the market return, which historically averages 12%, and a margin of (say) 3% to account for risk.

Figure 7.9 shows a sample calculation based upon a company borrowing rate of 8%, an expected return on investment of 15%, company debt of $50 million and equity of $35 million.

Although it is convenient to have a methodology that does suggest a rational discount rate, there is a disadvantage of discounting in this way. The WACC calculation suggests that the company's cost of capital is 10.9%. However, there is no reason to suppose that any project being evaluated would bear the same risk as the company. Also, there is no reason to suppose that the project would be financed in the same way as the rest of the company. Another frequent debate is whether different discount values should be applied for different projects. This is not advisable, as it would privilege one project over another and may even change the ranking of competing projects.

The APV method of discounting   The adjusted present value (APV) method recognises that cash flows carry different risks and should therefore be assessed using different discount rates to reflect this. The usual approach is to apply one discount rate equivalent to the cost of debt (R_d), and a second equivalent to the cost of equity (R_e). The most appropriate way of seeing the difference between WACC and APV methods is through an example, which focuses on purely operating costs and revenues.

Example of WACC discounting   In Figure 7.10 we can see a single year's operating results for a fictitious airline. The operating profit for the purpose of calculating tax is obtained by the operating cash flow and deducting the non-cash item of depreciation, as well as interest payments. Once the tax has been removed, the depreciation and

R_d = 8%
R_e = 15%
D = $50 million
E = $35 million

$$\text{WACC} = 0.08 \times \frac{50}{(50+35)} + 0.15 \times \frac{35}{(50+35)} = 10.9\%$$

Figure 7.9   Example of WACC Calculation

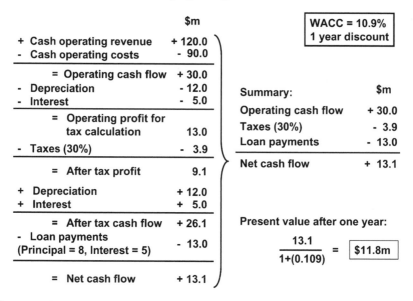

Figure 7.10   Cash Flow – WACC

interest are added back to give the after-tax cash flow. Loan payments are then further removed to arrive at the net cash flow, of $13.1m. It is this value which is discounted.

Example of APV discounting   There are good reasons why we should apply different discounting rates to different elements in the statement. Firstly, loan payment risk is represented by the bank's borrowing rate, which is lower than the company rate because the aircraft acts as security. In any case it is not strictly correct to discount loan repayments at a higher rate than the loan rate as this leads to a distortion of the present values. Tax is also discounted at the borrowing rate because the risk of a tax shield not materialising is the same as not paying the loan. In other words, if you do not repay the loan, you do not own the aircraft any more, so there is no tax shield. Operating cash flows carry more risk, so should be discounted at a higher discount rate, which is the R_e.

In the example in Figure 7.11 tax is applied on the full operating cash flow of $30m and then a tax 'shield' is introduced for both depreciation and interest. The separation is important because these two elements will bear a different discount rate. So, the operating cash flow less tax is discounted at the cost of equity (R_e), and the tax shields and the loan payments are discounted at the borrowing rate (R_d). The effect has been to reduce the present value for Year 1 from $11.8m to $11.0m.

The Internal Rate of Return

This second method using the time value of money is no more than a mathematical manipulation of something that we have already calculated. For a given discount rate there will be an NPV of a future series of cash flows that, hopefully, will be positive. As we have seen, increasing the discount rate decreases the present value of a future

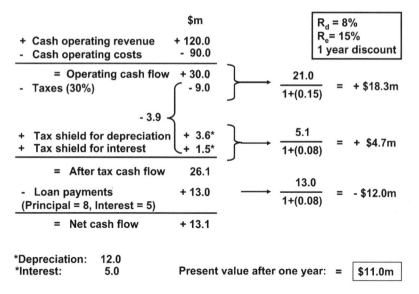

Figure 7.11   Cash Flow – APV

stream of cash flows. If we were to continue increasing the discount rate until the point where the net present value of all future cash flows would be equal to zero, then this discount rate would be the internal rate of return. It is found through a process of iteration. Although the IRR portrays a return on capital, it cannot distinguish between projects of varying size. In the eyes of many, the IRR is flawed because it can generate two results when cash flows oscillate between positive and negative values.

Choosing Between Competing Projects

On the face of it the project generating the highest NPV should be the most preferred, but it is important to visualise how NPV alters with discount rate, whether the hurdle rate is close to the cross-over between two projects, and how close the chosen project might be to generating a negative NPV. In Figure 7.12 Project B would appear to be the most satisfactory, but only at a hurdle rate of up to 10%. Beyond this point the ranking of the two projects changes. Even more interesting is that Project B would be a loss-maker at much higher discount rates. The decision would clearly need to embrace factors other than the NPV comparison.

Selecting the Right Time Period for a Cash Flow Study

Rather obviously, an NPV or IRR calculation can be altered simply by changing the length of time under study. However, choosing the right study period is fundamental to ensuring a fair comparison between two projects.

Imagine that we had to choose between two alternatives: one being to revamp and retain an elderly aircraft which has a remaining economic life of seven years; and secondly to replace that aircraft immediately with one which would have an

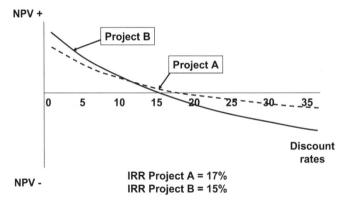

Figure 7.12   NPV Comparison

economic life of 20 years. It would be incorrect to compare the cash flows generated by these two alternatives for the unequal duration of their economic lives. One method would be to close the study at the end of the economic life of the old aircraft. In this case, we would need to reflect that the newer aircraft would still have 13 years' of use remaining, by 'imposing' a sale of the aircraft at that moment. An estimate of the market value of the aircraft would need to be written into the cash flow.

An alternative would be to carry out the study for the full 20 years' life of the new aircraft, but then we would need to make a presumption as to how we would replace the old, existing aircraft in seven years' time. This would require forecasting the price of the future replacement in seven years from the beginning of the study.

Neither of these two solutions is perfect, but it is vital to harmonise the economic lives of projects in order to eliminate bias.

Another decision, which must be made in setting up a cash flow, is how long each individual discounted time period should be. For most purposes, an annual present value should suffice. However, if the market is characterised by seasonal peaks and troughs, then we might wish to consider a bi-annual structure.

A cash flow for an operating lease study might warrant splitting the time zones into monthly components. This is because operating lease periods tend to be of shorter duration than if we were to purchase the aircraft. Also, the payments associated with the aircraft may vary a great deal month by month if, for example, the lease rate is linked to fluctuations in market interest rates, or the maintenance reserves are linked to changing utilisation levels.

Making the Choice

The great thing about using NPV or IRR is that we are presented with black and white choices. For competing projects that are independent in nature, one of them is going to generate a greater net present value.

Life is not, of course, that simple. Any NPV is highly contingent upon the discount rate, study period and forecasting of assumptions within. For this reason, scenario-building is essential. This can take the form of constructing an 'optimistic'

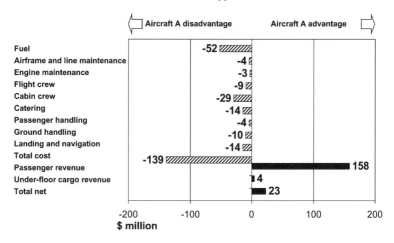

Figure 7.13 Incremental Present Value

scenario with favourable assumptions of economic growth and market share, stable costs and strong yields. Against this, we could test a 'pessimistic' scenario with opposing assumptions. Another method would be to test the impact of changes in key parameters on the results, such as fuel price. If this is done in conjunction with the prioritised list of key decision factors, then a more comprehensive picture can be built up.

In order to simplify the work of scenario building, combining probabilities of key inputs through a Monte Carlo simulation would more easily generate a set of NPVs for comparison. There are several commercial software packages that enable the tedious work of scenario building to be effectively done.

It can often be the case that an aircraft decision would impact so greatly on the operation of the business that it is almost impossible to embrace all of the relevant elements into the cash flow. In fact, a cash flow comparison of two aircraft alternatives is most likely to be an abbreviated picture. For this reason, it is sometimes appropriate to construct incremental cash flow studies, which are deliberately built to concentrate on key economic issues, without taking on board the entire picture.

An example of incremental cash flow modelling is seen in Figure 7.13. We can imagine that two aircraft, Aircraft A and Aircraft B, are being compared on the basis of their operating costs, with all ownership charges excluded. In the event of a discrepancy in seating configuration, it is important to clarify the basis on which the revenue estimation is made. Typically, we might assume that a percentage of the seating difference would be taken into account. Our fictitious example could assume 40% load factor for the seating delta, 50% cargo load factor, a cost escalation rate of 3%, and a revenue escalation rate of 2% (reflecting declining yield in real terms).

In Summary

In this Chapter we have seen that the investment appraisal can embrace a bigger picture than the aircraft operation itself. An appropriate aircraft selection cannot be made on the basis of aircraft economics alone. Many non-operating activities could be affected by the decision and a whole range of financial parameters must also be considered.

Crucially, the investment appraisal should be conducted so that the evolution of the project can be assessed over time. In other words, it is the movie that is important, rather than the snapshot.

Making the decision is not, of course, black and white. It is full of shades of grey, patterns and textures. The investment appraisal, essential though it is, can be no more than yet another aid to decision-making. Where one project is overwhelmingly better than another, then of course it would be easier to use the financial analysis to justify a decision. However, it is very often the case that alternatives are close to each other, complicating life. Where this happens, the fleet planner needs to be very clear about the assumptions, plus extraneous influences such as taxation and depreciation policies, for example. As we have seen, just a small change in the discount rate could swing the advantage one way or the other.

To complete our analysis of the process of fleet planning, and by way of a conclusion, we will now examine some of these additional complications in our quest to make that final decision, and take a broad look at some of the challenges that fleet planners will face in the future.

Chapter 8

Conclusion

Is Fleet Planning an Art or a Science?

An ideal fleet plan is akin to the Holy Grail. What we are trying to achieve is the impossible. We must solve a multi-dimensional problem where the conditions are changing almost every day. Resolving this problem requires lots of luck and a thick skin. We need the luck because the forecasting of market and economic conditions with sufficient accuracy for the long term is asking a great deal. We need the thick skin because fleet planners are easy targets for criticism when things go wrong.

There are many extraneous issues that can conspire to upset the cool logic of a calculated solution.

Who is Really Driving the Decision?

Countries that have a centralised planning structure are hamstrung when it comes to responding to rapidly changing market conditions. China's five-year planning system is obviously an additional constraint compared with a small company with an entrepreneurial spirit and run by a strong personality. Political interference can also be an unwelcome element to consider.

Another important issue of control concerns alliance strategies. Although alliances have now passed their embryonic stage, it is clear that there is a good deal of instability and partner-swapping going on. One can easily understand why airlines who are members of the strategic alliances are reluctant to commit to joint fleet plans as there is an ever-present risk of being left high and dry with a fleet mix that was designed to serve a different purpose. Alliance members continue to focus on revenue gain as a prime objective, with cost savings being so much more difficult to accomplish. It seems unlikely that the alliances will acquire aircraft in their own right owing to problems such as standard configurations and specifications, conflict in the timing of aircraft requirements among the alliance members, and a general reluctance on all sides to divulge confidential contractual conditions.

Regulatory and Infrastructure Issues

Slot availability and restrictions can have an overwhelming influence on capacity decisions, irrespective of what an airline would actually like to achieve in its market. For many years, a wholly artificial limit on the number of air transport movements at London Heathrow and Gatwick airports, due to environmental considerations, affected the way in which airlines planned their capacity. Regulators may also continue to require what they regard as dominant airlines to relinquish slots in exchange for anti-trust immunity, for example.

However, the appearance of the A380 on the scene has encouraged airlines to think carefully about the relationship between aircraft size and frequency at slot-constrained airports. Undoubtedly, many A380 customer decisions have been driven just as much by slot availability as by unit costs.

One country where there seems to be an imbalance between ordered capacity and slots is India. The severe infrastructure limitations seem at odds with the enormous amount of capacity ordered by Indian airlines in the 2000s.

The Manufacturer/Airline Relationship

Airlines that have always sourced their aircraft from one particular manufacturer may have a reluctance to switch to another supplier. This may be a matter of culture or simply a tendency to stick with what works. Making the break requires a considerably greater effort than to continue acquiring capacity from a tried and trusted supplier. Changing supplier may involve learning about a new aircraft technology as well as embracing a new culture and style of working. An extra degree of risk is associated with a fleet plan that necessitates a supplier switch. The same can be said for engine manufacturers and equipment vendors, too.

Pursuit of a dual-supplier strategy is a useful technique for an airline to extract full value from a deal. One reason why British Airways would not entertain the A340 is that the aircraft is offered without a choice of engine supplier. From an airline perspective, one can almost say that it is essential to generate and manage a competition between suppliers. Even if this is not possible, a 'competition' can be imagined between an existing fleet and a replacement option.

Are We Prisoners of History?

An airline is often bound by decisions made by earlier management teams that are difficult to unravel. Thus, Southwest Airlines' commitment to the 737 is unlikely ever to change. Changing earlier decisions may take a whole generation to accomplish. For example, SAS' decision to acquire A340 aircraft was taken after 11 years of 767 operation. South African Airways operated A320s for 12 years before deciding to revert to the 737-800. That decision was itself reversed and the A320 came back into the picture.

Will o' the wisp decisions can be costly in the long term. So, once a choice is made, there is often no economic prospect of changing for some time.

What Type of Airline are We Anyway?

This book has concentrated heavily on scheduled passenger airlines. However, the cargo, charter, leisure and low-cost businesses all exhibit different planning priorities and ways of doing things.

Although many of the analytical techniques are common, priorities may be very different. For example, a network comprising a collection of single sectors cannot be analysed in the same way as an integrated system channelling demand through a hub. Leisure and low-cost carriers have a greater focus on operational simplicity, whereas more traditional network carriers clearly continue to focus on product quality.

These differing strategies can mean that the fleet planning approach differs a great deal, even when we deal with the same aircraft type.

Lease vs. Purchase?

Operating leasing is a *means* of acquisition and is not strictly connected to the selection decision itself. More than one-quarter of the production of the two principal airframe manufacturers is destined for the leasing companies. Their criteria for buying aircraft are different from those of an airline. Mostly, they see the aircraft as an investment with a residual value potential.

Aircraft availability is of huge concern in most cases. The ability of the manufacturer to deliver a purchased aircraft to an airline is partly a function of the amount of his production already committed to the leasing fraternity. The manufacturer has an interest in regulating how many aircraft are in the hands of the leasing companies. Too many, and the manufacturer may risk competing with the lessors to place aircraft. Too few, and opportunities to place aircraft may be less evident.

One important function of the leasing companies is to place capacity in airlines that would not otherwise be in a position to purchase. Thus, the market coverage of aircraft types is broadened, perhaps paving the way from direct purchase at a later stage.

When does the Calculation Stop?

Much of this book has been devoted to setting out a structure for compiling a fleet plan and building up a dossier based upon quantifying performance and economics.

However, fleet planning decisions are being based more and more on intangibles such as trust and relationships with people. Developing the right chemistry between the parties in the deal is of equal importance.

It is very often the case that one aircraft will have an advantage over another in an area very difficult to calculate. We cannot place a monetary value on everything. Sometimes, intuition tells us that a certain solution is better than something else. Indeed, injecting too much subjectivity into what should be a calculated plan can actually devalue the result. Yet there is most certainly a place for such arguments, and they should be considered alongside the cold numbers.

In reality, once the numbers have been crunched, emphasis has to shift to other fundamental areas that all contribute to the business case. The list of items to take into consideration is endless, but may typically include:

Richness of the aircraft specification	Does one aircraft offer better value than the competitor?
Synergies with existing fleet	Are there savings in the number of spare engines, training or ground support equipment?

Who does the maintenance	Would the aircraft selection affect the choice, availability or cost of the maintenance provider?
ETOPS	Is it an opportunity or a burden?
Payload capacity	How closely does the seating and underfloor capacity match the market, both today and in the future?
Politics	Could this become an overriding factor? Can you really trust your supplier?
Timing	Is it the right time to invest at all? What about the availability of aircraft?

It is not always possible to assign a dollar value to the above items, but it is certainly not an option to ignore these issues.

What Should the Airline Expect from the Manufacturer?

In buying or acquiring an aircraft it is important to be assured that the supplier's business, product and market outlook are sound. Certain questions need to be raised, both internally and directly with the supplier. It is good practice to build a list of expectations of the supplier, which may be categorised as follows.

Firstly, it is reasonable to expect the manufacturer to demonstrate clearly that his product philosophy is economically competitive, both today and in the future. This means addressing both the cost and revenue sides of the equation. The product should also have the potential to grow and offer value and efficiency at all phases of its life. An ability to convert the aircraft at the end of its primary life into another use, such as a freighter, would obviously be important.

Secondly, it goes without saying that the product should comply with all certification requirements, with sufficient margin to continue to comply with future developments. This also goes for environmental constraints as well.

Thirdly, the manufacturer should be able to prove to the airline that he is competent in producing an aircraft on-time and with a high level of reliability in terms of both fuel burn, maintenance cost, operating weight empty, and dispatch reliability.

Lastly, it should be expected that the manufacturer has a clear technology road-map, so that the airline can be confident that innovation will continue to flow into the product line.

Some Crystal-Ball Gazing

This book has attempted to lay out the basic principles of fleet planning. The point has been made many times that the airline business is highly dynamic and in a state of constant change. That is what makes it so interesting. To bring the story to a conclusion I am going to put my head on the block and set out some key fleet planning issues that I personally believe will emerge in the future. There are five fundamental issues.

Firstly, we shall see an acceleration of the commoditisation of air travel. The low-cost carriers have led the way in forging a new philosophy that short-haul travellers have become primarily motivated by price, rather than service. The knock-on effect of this is that the traditional carriers are being forced to adapt to the new situation, and Aer Lingus is a perfect example of how a former flag-carrier has gone through a metamorphosis to emerge as a 'lower-cost carrier'. Having said that, it is unlikely that we shall see the commoditisation of the long-haul market. The impact of the low-cost revolution on aircraft configuration has yet to be fully felt, but it seems quite likely that a drive toward more standard aircraft specification will come about. This will be of benefit to the airlines, manufacturers, leasing companies and, ultimately, the passengers.

The second major change I envisage is that security concerns will assume a degree of such importance that the industry will be forced into radical change concerning the carriage of passengers' hand luggage. We saw that the industry has already gone through one change as a result of security concerns in the wake of the 11 September 2001 terrorist attacks, when strengthening of cockpit doors became mandatory. The emergence of the threat of explosives being smuggled on board aircraft in 2006 will very likely spawn a complete review of hand-luggage screening and, importantly, whether hand luggage should be physically separated from passengers. It is entirely possible that we shall see aircraft configurations that enable hand luggage accommodated in secure areas of the main cabin, accessed through a special procedure. Such a move would have an effect on the use of space of the aircraft and, possibly, the number of cabin crew (if dedicated security attendants become mandatory). Whatever happens, we can expect more pressure on aircraft operating costs due to security concerns.

We shall see an interesting interplay between the low-cost carriers attempting to outlaw all check-in baggage, and security concerns that push airlines to do the exact opposite. Either way, some lucrative opportunities for the manufacturers of minimalist luggage must surely emerge.

Growing concerns over security, more limitations on baggage and longer airport processing times is likely to give a boost to niche carriers operating premium services from less congested airports. The likes of Eos and MAXjet may find themselves with a willing market for their products.

One of the most serious risks in the longer term is whether the never-ending cycle of terrorism may result in an irreversible dampening of the rate of growth of air travel demand. The growth of discretionary travel could certainly be reduced if airport hassles become an epidemic.

Thirdly, over the next decade or so we shall move into the end-game concerning the debate on large vs. mid-sized aircraft for long-haul markets. Both the manufacturers have played their hands and it is now just a matter of time before judgement can be made as to who got it right. All indications suggest that the result will be a draw.

Fourthly, we will observe a fundamental shift in the centre of gravity in the airline business from the traditional centres of North America and Europe towards China, India and the Middle East. China is forecast by many to become the world's leading economic power by 2040. India's middle-class will certainly emerge to feed the burgeoning airline market. The Middle East hubs will continue to grow, with

Dubai in particular assuming a new role as the crossroads of the world. It is possible to fly from Dubai in any direction for 20 hours and reach any airport on the planet non-stop. Now that aircraft are available with such capability, this region will not hesitate to exploit these opportunities.

Lastly, the big challenge for the aircraft manufacturers and the airlines will be in how to extract even more efficiency from the aircraft itself, to reflect growing expectations to limit our impact on the environment. In this book we have observed that it is becoming increasingly challenging to improve aircraft performance. One key element will continue to be commonality and minimising the number of aircraft types in a fleet. Another tendency could be for airlines to design their networks in a more homogenous and efficient manner, thereby avoiding the trap of needing complex fleets of aircraft.

Rather like athletes attempting to break a new record, we are approaching the limits of what is physically achieved. So, attention will be turned elsewhere. We can expect more focus on efficient network design, in order to boost load factors and aircraft utilisation, and the manufacturers will come under more pressure to improve reliability and reduce delivery lead-times. And so the wheel will turn full circle. These goals can only be achieved through simplification and commoditisation of the product.

When I worked in the Airbus Marketing organisation, people would often come into my office and ask me for help with a fleet planning problem. They would plead with me to provide an answer. I would tell them all that there is no such thing as a silver bullet. In other words, there is never a simple, all-embracing solution to a complex problem. Every single case is unique and deserves its own tailored solution. Yet we still need to have a structure upon which we can build, based upon solid analytical principles. That is what this book has attempted to provide.

The real decision, however, comes from the heart and not the numbers. But no matter what the heart tells us, Robbie Burns' prediction, that the best laid plans often go wrong, often comes true!

Index